NIST Technical Note 1458

NIST Measurement Service for DC Standard Resistors

Randolph E. Elmquist
Dean G. Jarrett
George R. Jones, Jr.
Marlin E. Kraft
Scott H. Shields
Ronald F. Dziuba

Electronics and Electrical Engineering Laboratory
National Institute of Standards and Technology
Gaithersburg, MD 20899

December 2003

U.S. Department of Commerce
Donald L. Evans, Secretary

Technology Administration
Phillip J. Bond, Under Secretary for Technology

National Institute of Standards and Technology
Arden L. Bement, Jr., Director

CONTENTS

1. INTRODUCTION 1
2. DESCRIPTION OF SERVICE 2
   2.1 Special Standard Resistors (1 Ω and 10 kΩ) 2
   2.2 Standard Resistors (10^{-4} Ω to 10^6 Ω) 3
   2.3 High-Value Standard Resistors (10^7 Ω to 10^{12} Ω) 3
   2.4 Standard Resistors for Current Measurements – Shunts 4
   2.5 Resistance MAP Service 4
   2.6 Special Resistance Measurements 4
3. MEASUREMENT METHODS 5
4. DIRECT CURRENT COMPARATOR SYSTEMS 6
   4.1 Theory of Operation 6
   4.2 DCC Potentiometer System for 1 Ω Measurements 9
      4.2.1 Description of System 9
   4.3 DCC Potentiometer for 10 Ω and 100 Ω Measurements 12
      4.3.1 Description of System 12
   4.4 Manual DCC Resistance Bridge 13
      4.4.1 Description of System 13
      4.4.2 Measurement of High-Current Standard Resistors 14
   4.5 Automated Binary DCC Bridge 17
      4.5.1 Measurement Sensitivity 17
      4.5.2 DCC System Software Description 17
5. RESISTANCE-RATIO BRIDGES 19
   5.1 Theory of Operation 20
   5.2 Automatic Guarded Warshawsky Bridge 22
6. RING METHOD 25
   6.1 Theory of Operation 25
      6.1.1 First Subset of Voltage Measurements 25
      6.1.2 Second Subset of Voltage Measurements 27
      6.1.3 Third Subset of Voltage Measurements 28
   6.2 Data Analysis 28
   6.3 Guard Network 29
7. HIGH RESISTANCE MEASUREMENTS 30
   7.1 High-Resistance Standards 30
   7.2 Guarded Active-Arm Bridge 31
   7.3 Measurement Process 33
      7.3.1 Balancing Algorithm 33
      7.3.2 Automatic Resistor Selection 33
8. THE U.S. REPRESENTATION OF THE OHM 34
   8.1 Quantized Hall Resistance 34
   8.2 Primary Scaling 35
   8.3 CCC Ratios 36
   8.4 DCC ratios 38
   8.5 Transfer Standards 38
   8.6 Active-Arm Bridge Ratios 42

9. MEASUREMENT UNCERTAINTY 42
   9.1 Model - Standard Resistors 42
   9.2 Model - Measurement Systems 44
   9.3 Type A Standard Uncertainty 47
   9.4 Type B Standard Uncertainty... 47
      9.4.1 Standard Resistors... 47
      9.4.2 Measurement Systems 49
      9.4.3 Measurement Repeatability.. 53
   9.5 Combined Uncertainty... 54
   9.6 Reported Uncertainty 54
   9.7 Current Shunt Uncertainties... 57
10. QUALITY CONTROL... 59
11. ONGOING PROJECTS 60
12. ACKNOWLEDGMENTS 61
13. REFERENCES 61
14. BIBLIOGRAPHY 64
15. APPENDIX... 65
   15.1 Report for a 1 Ω "Thomas-type" Standard Resistor 65
   15.2 Report for a 10 kΩ "Special Air-type" Standard Resistor.. 65
   15.3 Report for a High-Value Standard Resistor 65

NIST MEASUREMENT SERVICE FOR DC STANDARD RESISTORS

Randolph E. Elmquist, Dean G. Jarrett, George R. Jones, Jr.,
Marlin E. Kraft, Scott H. Shields, and Ronald F. Dziuba

Abstract - At the National Institute of Standards and Technology (NIST), the U.S. representation of the ohm is based on the quantum Hall effect, and it is maintained and disseminated at various resistance levels by working reference groups of standards. This document describes the measurement systems and procedures used to calibrate standard resistors and current shunts of nominal decade values in the resistance range from 10^{-5} Ω to 10^{12} Ω. Resistance scaling techniques used to assign values to the working standards are discussed. Also included is an assessment of the calibration uncertainties at each resistance level.

1. INTRODUCTION

Since January 1, 1990, the U.S. representation of the ohm has been based on the quantum Hall effect in which a resistance is related to the ratio of fundamental constants h/e^2 [1,2]. The quantized Hall resistance, R_H, is defined as the quotient of the Hall voltage V_H of the ith plateau to the current I in the Hall device and is given by

$$R_H(i) = V_H(i)/I = R_K/i,$$

where the von Klitzing constant R_K is believed to be equal to h/e^2, and i is an integer of the quantum Hall state. The value of the U. S. representation of the ohm is consistent with the conventional value of the von Klitzing constant, i. e.,

$$R_{K-90} = 25\,812.807\text{ Ω},$$

exactly, adopted internationally for use in representing the ohm beginning January 1, 1990. This conventional value is believed to be consistent with the SI ohm to within 0.005 Ω which is the value assigned to be the combined standard uncertainty, corresponding to a relative uncertainty of 0.2 μΩ/Ω (0.2 ppm)*. Because this uncertainty is the same for all national laboratories and is not relevant for traceability to national standards, it is not included as a component of the uncertainties given in NIST Reports of Calibration for standard resistors. However, its existence must be taken into account when the utmost consistency between electrical and non-electrical measurements of the same physical quantity is required.

For precision measurements using the quantum Hall effect, the integer i is usually chosen to be either 2 or 4, resulting in quantized Hall resistances of 12 906.4035 Ω or 6453.20175 Ω [3]. These odd resistance values, along with the complexity of the experiment, do not lend

---

* The abbreviation ppm (parts-per-million) is used in place of an equivalent SI ratio such as μΩ/Ω or μV/V when it does not appear in combination with SI terms.

themselves to the routine support of the calibration of standard resistors of nominal decade values. Therefore, at NIST the ohm has continued to be maintained for calibration purposes using groups of standard resistors whose predicted mean value is checked periodically against the quantum Hall effect and adjusted, if necessary. These standard resistors have been fully characterized for drift, temperature, pressure, and load coefficients.

NIST provides a calibration service for standard resistors of nominal decade values (i.e., $R = 10^n$ where n is an integer) in the range between $10^{-5}\,\Omega$ and $10^{12}\,\Omega$. In addition, non-decade value standard resistors can be calibrated in the lower resistance ranges, near $10^3\,\Omega$ or below. To provide this wide-ranging calibration service, NIST maintains a working reference group at each nominal decade value between $1\,\Omega$ and $10^{12}\,\Omega$. The working reference groups are calibrated using special ratio techniques. A customer's resistor is calibrated against the NIST working reference group in a manual or an automated comparison measurement system using either a resistance-ratio bridge or a flux-balancing type of current comparator. The specific measurement system used depends upon the value of the resistor being calibrated and on the accuracy required or warranted. NIST procedures follow recognized dc measurement practices such as current reversal, minimum power dissipation consistent with desired resolution and accuracy, and the use of check standards and statistical data analysis techniques to monitor the operation of the measurement systems.

2. DESCRIPTION OF SERVICE

NIST provides a calibration service for low-current standard resistors with nominal decade values in the range between $10^{-4}\,\Omega$ and $10^{12}\,\Omega$. High-current standard resistors with nominal values as low as $10^{-5}\,\Omega$ can be calibrated at currents as high as 2000 A. In addition to this regular calibration service, NIST offers a resistance Measurement Assurance Program (MAP) service at the $1\,\Omega$ and $10\,k\Omega$ levels. Other special resistance measurements are undertaken if they require the unique capabilities of NIST.

The latest issue of the NIST Special Publication 250 entitled "NIST Calibration Services Users Guide" contains detailed descriptions of the currently available NIST calibration services [4]. It also contains information on the scheduling of these services along with recommended procedures for shipping a standard to NIST for calibration. A separate Fee Schedule, NIST Special Publication 250 Appendix, lists the costs of services, and it is updated periodically to reflect changes in prices and services.

2.1 Special Standard Resistors (1 Ω and 10 kΩ)

NIST recognizes a special category of $1\,\Omega$ and $10\,k\Omega$ standard resistors which exhibit small corrections of < 10 ppm from nominal value, high stability, and a low temperature coefficient of resistance (TCR); accordingly, these resistors merit the best possible measurement uncertainties. In this category are the Thomas-type $1\,\Omega$ resistors, or their equivalent, typically having drift rates of $<\pm 0.1$ ($\mu\Omega/\Omega$)/year and TCR's of (0 ± 3) ($\mu\Omega/\Omega$)/K, and the Evanohm[*]-type 10 kΩ resistors having drift rates of $<\pm 0.2$ ($\mu\Omega/\Omega$)/year and TCR's of (0 ± 1) ($\mu\Omega/\Omega$)/K. Measurement parameters of temperature and current for these resistors are given in Table 2.1.

These resistors are acclimatized in their respective test environment for approximately one week prior to calibration. The temperature of the resistor at the time of the measurement is given in the Report of Calibration. Since some resistors have a significant pressure coefficient, the pressure at the time of the measurement is also reported for these 1 Ω and 10 kΩ standards.

Table 2.1. Measurement parameters for special standard resistors

Resistor	Medium	Temperature	Current
1 Ω	oil	(25.000 ± 0.003) °C	100 mA
10 kΩ	oil	(25.00 ± 0.01) °C	1 mA
10 kΩ	air	(23.0 ± 1.0) °C	1 mA

2.2 Standard Resistors (10^{-4} Ω to 10^6 Ω)

Standard resistors with nominal decade values in the range between 10^{-4} Ω and 10^6 Ω with fixed terminations and permanently identifying serial numbers are accepted for calibration. Non-decade value standard resistors of similar quality can also be calibrated if the value is less than 1 kΩ. In general, these resistors are characterized by 1) TCR's of (0 ± 10) ($\mu\Omega/\Omega$)/K at the temperature of use, and 2) drift rates of less than ± 5 ($\mu\Omega/\Omega$)/year. Usually, these standard resistors are measured in an oil bath maintained at a temperature of (25.00 ± 0.01) °C, and at a power dissipation level of 10 mW. Resistors permanently mounted in temperature-controlled enclosures, or designed for operation in an air environment near 23 °C are also accepted for calibration. At the levels of uncertainties reported, four-terminal connections are required for standard resistors of nominal values 10 kΩ and below.

2.3 High-Value Standard Resistors (10^7 Ω to 10^{12} Ω)

High-value standard resistors in the range between 10^7 Ω and 10^{12} Ω are calibrated in an air bath maintained at a temperature of (23 ± 0.1)°C and at a relative humidity of (35 ± 5)%. The maximum test voltage is 500 V for nominal resistor values $< 10^{10}$ Ω, and 1000 V for resistor values $= 10^{10}$ Ω. Only resistors that are mounted in a shielded enclosure, with permanently identifying serial numbers and with suitable terminations, are accepted for calibration. The resistance of a film-type standard resistor is frequently highly voltage dependent. Hence, the magnitude of the test voltage should be specified when this type of resistor is submitted for

---

[*]Evanohm is a commercial alloy having a resistivity of about 1.34 $\mu\Omega$-m with a nominal composition of 75 % Ni, 20 % Cr, 2.5 % Cu, and 2.5 % Al. By suitable annealing and heat treatment, its TCR can be adjusted to nearly zero from (20 to 30) °C.

calibration. The temperature, relative humidity, and test voltage of the resistor are given in the NIST Report of Calibration.

2.4 Standard Resistors for Current Measurements – Shunts

Four-terminal standard resistors for use in the precise measurement of high direct current are calibrated by comparison with NIST working standards. These typically are 1) individual units with resistance values in the range between $10^{-5}\,\Omega$ and $10^{-1}\,\Omega$, or 2) multi-value or multi-range shunts, typically with nine resistors, for measurement of nine currents ranges from 10 µA to 300 A. The desired measurement current(s) should be specified, and should not exceed the rated current. The final resistance, after approximate thermal equilibrium is reached for the specified current, is reported in the Report of Calibration. All current shunt standards are measured in the NIST laboratory air environment and the air temperature is measured and reported. If desired, NIST will also report the equilibrium temperature of a copper-constantan thermocouple permanently attached to the resistance element.

2.5 Resistance MAP Service

Resistance MAP transfers [5] are offered at the $1\,\Omega$ and $10\,k\Omega$ resistance levels. Four well-characterized standard resistors are used as transport standards in each transfer. The suggested measurement schedule at the customer's laboratory consists of measurements on each transport resistor three times a week for a period of 4 to 6 weeks depending upon the settling time of the resistors due to effects caused by transportation.

Participation in this program is generally not advisable unless a laboratory is 1) required to support resistance measurements at or near the state-of-the-art in accuracy, and 2) willing to adopt a system for the continuous surveillance of standards during the intervals between MAP transfers. A successful transfer requires a considerable amount of data collection and a willingness to become involved in the data analysis process. Data supplied during routine NIST calibrations suffice for normal measurement requirements of standards laboratories if proper methods are used by the laboratory to quantify the additional uncertainties caused by transportation and their own measurement process.

2.6 Special Resistance Measurements

Special resistance measurements such as the determination of the temperature coefficient, (power) load coefficient, or pressure coefficient of resistance of a standard resistor; testing or evaluation of prototype state-of-the-art resistance standards; high-value resistance measurements above the $10^{12}\,\Omega$ level; and other calibration services not specified above are made at the discretion of the NIST technical staff by prearrangement. Also offered as a special test service is the calibration of standard resistors using the quantum Hall effect by direct comparison or through a MAP method internal to NIST. These special tests are offered at certain times by prearrangement and only for high-quality standards at the $1\,\Omega$, $100\,\Omega$, $10\,k\Omega$, and $1\,M\Omega$ levels, and for suitable standards having values equal to the nominal quantized Hall resistance (QHR) at the $12\,906.4\,\Omega$ level. Contact NIST staff listed in the Internet address below for more details:
http://www.ts.nist.gov/ts/htdocs/230/233/calibrations/Electromagnetic/Resistance.htm

3. MEASUREMENT METHODS

To provide a wide-ranging calibration service at NIST for 18 decades of resistance from 10^{-5} Ω to 10^{12} Ω requires the application of four basic methods that are used in seven stand-alone measurement systems for comparing standard resistors. Listed below are the seven measurement systems along with their primary resistance measurement levels.

1) Automated current comparators for special 1 Ω measurements.
2) Automated current comparator for 100 Ω measurements.
3) Automated current comparator with range extender for measurements $\leq$ 1 kΩ.
4) Manual current comparator with range extender for high-current measurements.
5) Automated unbalanced bridge for 1 kΩ to 1 MΩ measurements.
6) Automated guarded Warshawsky bridge for special 10 kΩ and 1 MΩ measurements.
7) Automated active-arm bridge system for 10^7 Ω to 10^{12} Ω measurements.

Although each system is unique, they share all or some of the following measurement concepts and techniques:

1) Working Standards - NIST standard resistors, whose values are based on the quantum Hall effect. The mean value of one or more working standards is used to assign a value to a resistor under test. The drift in value of NIST working standards is in general highly predictable using a linear model, based on prior observations and careful maintenance of the resistors and their environment. For the 1 Ω working group, and generally for other working standards at NIST, a linear equation fitting the scaling data is used to calculate predicted reference values. The coefficients of the linear equation that best fits recent scaling data are the *drift factors* of that particular working standard.

2) Check Standard - A NIST standard resistor that is treated as an unknown and measured during each measurement run. For any given measurement run, the difference between the measured and predicted values for a check standard indicates whether the measurement system is operating under statistical control.

3) Dummy Resistor - A resistor, of equal quality to a working standard resistor, used as a short-term reference in a bridge or current comparator measurement system. Its absolute value need not be known; however, it must remain stable or predictable during a measurement run.

4) Substitution Technique - When the working standards and test resistors of the same nominal value are indirectly compared by substitution in the same positional arm of a bridge circuit. This technique tends to cancel errors resulting from ratio non-linearity, leakage currents, and lead and contact resistances. Also, comparing the measured and predicted differences among NIST working standards contributes to the analysis of measurement uncertainty.

5) Reversal Technique – When a null detector balance is achieved for each current polarity, upon reversal of the battery or power supply connections to the circuit. This eliminates the effect of

constant thermal electro-motive force (EMF) in the detector circuit. With at least two reversals, the effect of slowly varying thermal EMF also can be eliminated.

4. DIRECT CURRENT COMPARATOR SYSTEMS

Direct current comparator (DCC) systems [6-8] are used at NIST to calibrate four-terminal standard resistors up to values around the 1 kΩ level. The DCC, because of its high level of resolution, insensitivity to lead resistances, excellent ratio linearity and ratio stability, has replaced resistance-ratio instruments such as the Wenner and Kelvin bridges for these resistance measurements. Another distinguishing feature of the DCC in contrast to resistance-ratio bridges is that the balance condition is obtained under equal voltage drops across the resistors. For ratios other than 1-to-1, this results in the greater power dissipation occurring in the lower-value resistor which is usually designed to withstand higher currents. A commercial manual DCC resistance bridge can be used to calibrate high-current standard resistors at NIST, at currents up to 2000 A, with the reference standard dissipating no more than 10 mW.

In 1969, a DCC potentiometer system replaced the NIST Wenner bridge for the intercomparison of Thomas-type 1 Ω standard resistors. With the improved precision of this system, a correlation of the change in resistance of a Thomas-type resistor with variations of barometric pressure was detected. This led to the determination of the pressure coefficients of the NIST set of Thomas-type standard resistors in order to correct for this effect. The magnitude of the ambient pressure is now included in the calibration report for a Thomas-type resistor. This NIST DCC potentiometer system for the calibration of 1 Ω standard resistors was automated in 1982 [9]. In 1984, a second DCC potentiometer was modified for the intercomparison of 10 Ω and 100 Ω standard resistors. This system was automated in 1985.

Newer NIST DCC systems are commercial automated binary DCC bridges purchased in 1996 and 2003, and are used for many calibration measurements for resistance values up to 10 kΩ, and for high-current resistors. The 1996 system consists of an automated current comparator bridge with internal 150 mA sources, a 100 A range extender unit, a scanner with selecting relays for the resistance standards, a 0 A to 100 A power supply, and a 0 A to 5 A power supply. The 2003 DCC system contains updated electronics and extends the current range by an additional step to 400 A. This 400 A DCC system is undergoing installation and testing at the time of publication.

4.1 Theory of Operation

In general, the operation of a current comparator is based on Ampere's fundamental law

$$\oint H \cdot dl = \Sigma I \quad ,$$

where the line integral of the magnetic field H around a closed path dl is equal to the total current I crossing any surface bounded by this path. For an ideal comparator at balance, the total current is carried by two ratio windings (primary and secondary) and the total number of ampere-turns of one winding is equal and opposite to that of the other winding; thus,

$$\Sigma I = N_P I_P - N_S I_S = 0 \quad ,$$

where the subscripts P and S refer to the primary and secondary windings, respectively. The current ratio of the comparator, I_P/I_S, is therefore equal to the inverse of the turns ratio, N_S/N_P. To obtain high ratio accuracies, the ampere-turn balance or zero flux condition is determined by some type of flux detector system that is only sensitive to the mutual fluxes generated by the ratio windings.

In practice, a DCC achieves good ratio accuracy and sensitivity by utilizing high-permeability toroidal cores, magnetic and eddy-current shields, and careful winding procedures. The main component of a DCC consists of a pair of high-permeability cores, surrounded by a magnetic shield, over which are the ratio windings that carry the direct currents to be compared. Cryogenic current comparators are similar in concept, but make use of the ideal magnetic shielding properties of a self-enclosing, but non-continuous, surface made of superconducting material.

The currents for the ratio windings are supplied by two isolated direct current sources. Around the cores and within the magnetic shield of the DCC is wound a modulation-detection winding that is used to sense the flux condition of the cores. This is achieved by modulating the core permeability and using a second harmonic detector circuit. The presence of dc flux in the cores due to primary and secondary ampere-turn imbalance is indicated by this detector output both in magnitude and polarity. The detector output is used in a feedback circuit to adjust the current in one of the windings, automatically maintaining ampere-turn balance.

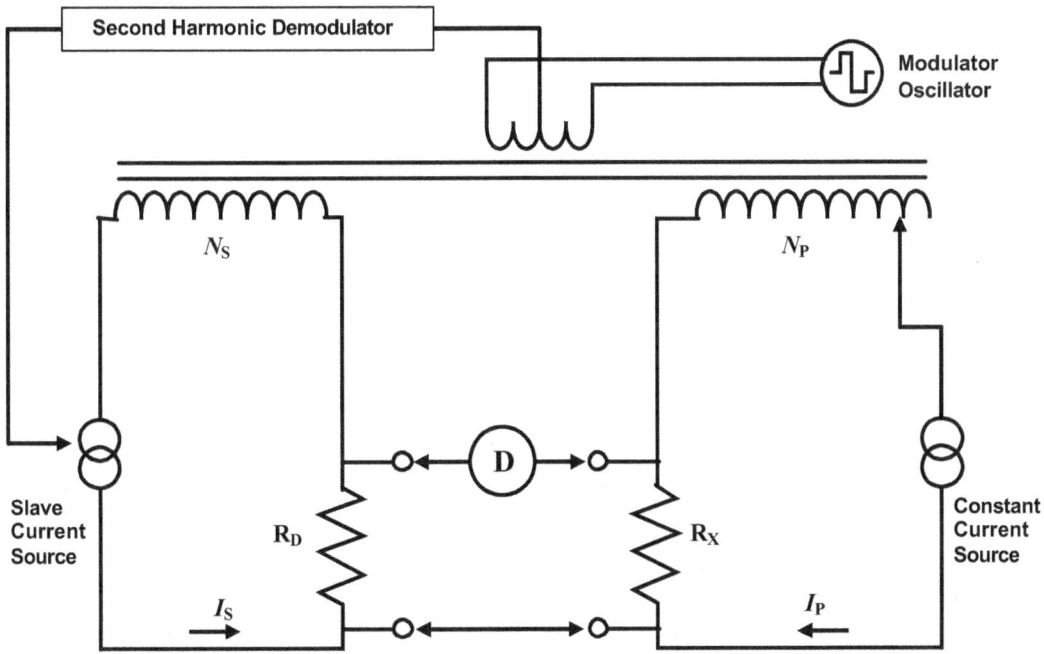

Fig. 4.1. Basic circuit of a DCC

The basic circuit of a self-balancing DCC resistance bridge is shown in Fig. 4.1. This bridge requires two simultaneous balances, an ampere-turn balance and a voltage balance. The slave current source is continuously adjusted in a feedback mode by the output of the demodulator circuit so that ampere-turn balance is maintained. Under this condition, the ampere-turn product of the primary circuit equals that of the secondary circuit, i.e.,

$$N_P I_P = N_S I_S \quad . \tag{4.1}$$

The voltage balance can be achieved by the adjustment of the number of turns in the primary circuit, N_P, until there is a null condition on detector D. Then the voltage drop across the unknown resistor R_X[**] in the primary circuit is equal to the voltage drop across the dummy resistor R_D in the secondary circuit or,

$$R_X I_P = R_D I_S ,$$

and using eq (4.1), the value of R_x can be expressed as

$$R_X = (N_P / N_S) R_D \quad . \tag{4.2}$$

Similarly, another measurement with the direct substitution of a known standard resistor R_S of the same nominal value in the primary circuit and re-balancing the detector by adjusting N_P to a new value N'_P results in

$$R_S = (N'_P / N_S) R_D \quad . \tag{4.3}$$

Combining eqs (4.2 and 4.3) gives

$$R_X = \left(1 + \frac{\Delta N_P}{N'_P}\right) R_S , \tag{4.4}$$

where ΔN_P is the difference in the number of primary turns ($N_P - N'_P$) between the two voltage balance settings.

Thus, the value of an unknown resistor R_X is determined in terms of a standard resistor and the small relative difference in turns ratio. The dummy resistor R_D must remain stable during the time interval the two voltage balances are made; otherwise an error will occur in the measurements. For the DCC potentiometer systems which include the $1\,\Omega$ and $100\,\Omega$ measurements, a string of working standard and test resistors are connected in series in the primary circuit. Then each resistor in turn is compared to the dummy resistor. If more than one working standard resistor is used in this measurement process, then the value of an unknown resistor is based on the mean value of the working group.

[**]Roman or upright type is used to denote a component (e.g., R for resistor, C for capacitor) and italic type is used to represent its physical quantity (e.g., R for resistance, C for capacitance).

4.2 DCC Potentiometer System for 1 Ω Measurements

4.2.1 Description of System

A schematic diagram of the DCC potentiometer circuit for 1 Ω measurements is shown in Fig. 4.2. Fifteen Thomas-type resistors or their equivalent - five of which comprise the working group, along with two check standards and eight unknown resistors - are connected in series within the primary circuit of the DCC. The value of any resistor in the string can be determined by indirectly comparing its voltage drop to the mean of the voltage drops of the reference group via a stable, 0.5 Ω dummy resistor (R_D) in the secondary circuit of the DCC. The resistors are mounted on a mercury stand that is housed in a constant temperature oil bath controlled at (25.000 ± 0.003) °C. The voltage balance is made automatically by driving an isolated electronic nanovoltmeter detector (D) to a null condition with a feedback current through an auxiliary 10 turn winding. This feedback current, which is proportional to the difference between the test and dummy resistors, is monitored by measuring the voltage drop across a 100 Ω resistor in series with the auxiliary winding using a digital nanovoltmeter (DVM). The feedback circuit is calibrated by inserting an additional unit turn in the primary circuit of the DCC which changes the voltage balance by exactly 500 ppm. A personal computer (PC) controls the operation of the DVM, resistor selection, current reversal, and the insertion of the feedback calibration signal. The PC also monitors the oil bath temperature, ambient temperature, ambient relative humidity, and barometric pressure.

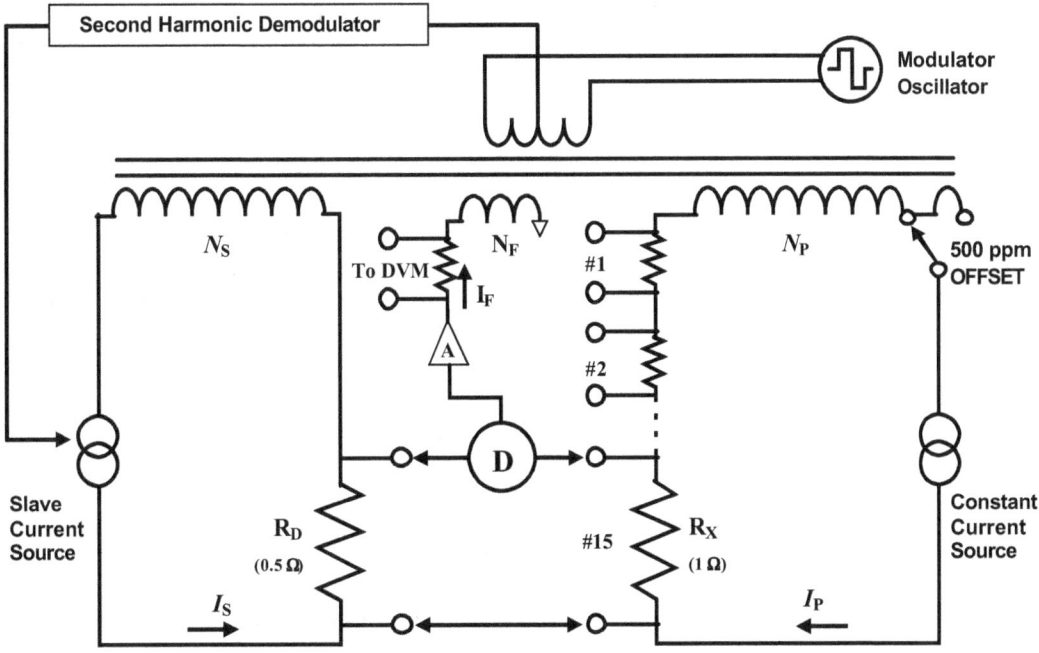

Fig. 4.2. Automated DCC potentiometer for 1 Ω measurements.

A description of some of the important features of the main components of the system follows:

1. DCC Potentiometer

The DCC is a commercial unit having an adjustable 2000-turn primary winding, a 1000-turn secondary winding, and a fixed 0.1 A primary current source. The system noise level is specified by the manufacturer to be less than 1 µA. The constant current source is specified as having a voltage compliance of 30 V, an output resistance of 10 GΩ, and a stability of 1 (µA/A)/day plus the instability of its voltage reference standard cell. To automate the system, four major modifications of the comparator were necessary:

A. The photodetector galvanometer amplifier (PGA) detector has been bypassed and a low-noise, isolated electronic nanovoltmeter detects the main voltage balance condition. Analog output from the nanovoltmeter is integrated and the resulting signal is buffered by an operational amplifier. This signal is passed through a 100 Ω resistor and the 10-turn auxiliary winding, driving the voltage sensed by the detector to a null condition.

B. The push-button current reversing switch was replaced with a relay. It is connected to a relay control module interfaced to the PC.

C. A DPDT relay with mercury-wetted contacts was connected to the 1 turn/step switch of the DCC between the "0" and "1" dial positions. This relay is also connected to the relay control module. When energized this relay provides a 500 ppm bridge offset.

D. The current divider circuit used for the "standardize" mode of the DCC was disconnected from its 10 turn winding. This winding is used in the feedback circuit which provides for an automatic voltage balance.

2. Feedback Circuit

The isolated nanovoltmeter detector (D) provides an isolated output signal that is integrated and buffered to produce a feedback current I_F through the 10 turn winding N_F. The current I_F is monitored by measuring the voltage drop across a 100 Ω resistor with the nanovolt DVM, and is a measure of the difference in resistance corrections for the calibration ratio of 2-to-1.

In order to determine the detector system sensitivity, a single-turn offset is added to the 2000 turn winding, and this ratio becomes 2001-to-1000. The output of the DCC bridge demodulator circuit automatically increases the 200 mA secondary current by exactly 500 ppm (0.1 mA) to compensate for the change in ampere-turns in the primary winding. The corresponding voltage change of 500 ppm across the 0.5 Ω dummy resistor is cancelled by the voltage feedback circuit. To affect this cancellation, a current change of 10 mA is produced in the 10 turn winding N_F and this current change is measured by the DVM.

3. Resistor Switch Module

A modified crossbar switch selects the appropriate potential terminals of a resistor in the primary circuit of the DCC whose voltage drop is to be compared to that of the dummy resistor in the secondary circuit. The switch mechanism is separated from the electromagnetic actuator coils and placed in a heavy aluminum box. The coils are mounted outside the box and their push rods are extended by plastic rods. This type of separation reduces heat transfer from coil to switch point, and the heavy aluminum box effectively eliminates thermal gradients within the box. The crossbar switch uses beryllium-copper conductors and gold-silver-platinum contacts which are thermoelectrically matched to each other resulting in thermal EMFs of less than 250 nV/K.

4. Resistance Thermometer Bridge

The temperature of the oil bath, maintained at (25.000 ± 0.003) °C, is monitored with a calibrated platinum resistance thermometer (PRT). The PRT is calibrated in terms of the International Temperature Scale of 1990 (ITS-90), and its resistance of about 28 Ω at the control temperature is measured using an automated resistance thermometer bridge [10]. The bridge has a range up to 101.1 Ω, a resolution of 1 µΩ, and an accuracy limited by the resolution or 0.1 ppm. The bridge has four input ports and is remotely controlled through an IEEE 488 bus. The bridge is calibrated prior to each PRT measurement against a known 100 Ω resistor. The PRT and resistor are measured at a current level of 1 mA.

5. Pressure Transducer

The barometric pressure is monitored using a calibrated commercial transducer based on a silicon microelectronic technique. This transducer consists of a micro-machined vacuum-gap silicon diaphragm that bends when barometric pressure changes. This changes the capacitance of the sensor, which is measured and converted into a pressure reading. The sensitivity of the transducer is 7 Pa (0.05 mm Hg) with an accuracy of 15 Pa (0.1 mm Hg). The accuracy of the transducer is periodically checked against a calibrated aneroid barometer.

6. Temperature-Humidity Indicator

Ambient temperature and relative humidity are monitored by a commercial digital thermometer/hygrometer instrument. The instrument has a resolution and reproducibility of 0.1 °C and 0.1 % RH. The temperature and relative humidity analog recorder outputs are measured by a DVM which is interfaced to the PC via the IEEE-488 bus.

7. Personal Computer

The PC is configured with hard and diskette drives for program and data storage. It contains an IEEE-488 board and a parallel I/O board to control the crossbar switch and DCC, and an RS-232 interface to obtain data from the pressure and humidity transmitter.

8. Software

The computer program for operating the system is written in BASIC language using multiple subroutines to handle the data taking and data processing. The data file of a test run is stored on the PC hard drive and on a floppy disk, and manually transferred by floppy disk to another PC with access to the local network server computer and NIST intranet. The drift factors and relevant characteristics of the five working standards and two check standards are stored in a remote NIST database, and in a file on the server computer. NIST has developed a program "THOMAS" that executes on the server computer to compute, analyze, and store the results of the test run. This program adjusts the predicted value of the group mean of the five reference resistors for variations of the oil bath temperature and ambient pressure from their respective nominal values of 25.000 °C and 101,325 Pa (760 mm Hg). This corrected group mean is then used to calculate the values of all of the unknown resistors in the measurement string for the test run including the values of the check standards. The values of NIST resistors corrected to a temperature of 25.000 °C and a pressure of 101,325 Pa are stored in individual data files according to their NIST serial number. The essential data for all resistors are saved in local files and in a NIST database. The data are saved in a format so that a linear regression fit of the data vs. time can be calculated using another system program. The THOMAS program also provides a check on the results by flagging the mean value of a resistor if its standard deviation exceeds a predetermined value (see section 10).

4.3 DCC Potentiometer for 10 Ω and 100 Ω Measurements

The design, construction, and operation of this system are similar to that for the automated 1 Ω system. The following description and operation sections will emphasize the differences between the two systems. Similarities between the two systems will be briefly mentioned. For more details on these similar features, refer to section 4.2 of this report.

4.3.1 Description of System

The DCC potentiometer circuit for 10 Ω and 100 Ω measurements resembles that shown in Fig. 4.2 for the automated 1 Ω system. However instead of 15 resistors, the system is designed to compare eight resistors of the same nominal value: two working standards, one check standard, and five unknown resistors. These resistors are mounted on a mercury stand housed in a constant temperature oil bath controlled at (25.000 ± 0.003) °C. The resistors for the secondary circuit are sealed in a heavy aluminum box filled with silicone-based oil and the box is immersed in the oil bath. The DCC is operated at a 2-to-1 ratio; hence this box contains a 5 Ω resistor and a 50 Ω resistor that are needed for measuring 10 Ω and 100 Ω resistors in the primary circuit, respectively.

In similar fashion to the 1 Ω system, the DCC was modified to provide 1) automatic current reversal, 2) automatic ratio offset, 3) an external access to a 10 turn winding, and 4) an isolated detector output circuit. The detector-feedback circuit is identical to the one used in the 1 Ω system except that the detector (D) gain setting is reduced to prevent oscillations from occurring in the system.

The original 50 mA constant current source (CCS) for the primary circuit of the DCC was modified to provide output currents of 10 mA and 31.6 mA. The excellent stability of the CCS is obtained by sending the load current through a reference resistor whose voltage drop is connected in series opposition to the reference voltage of an unsaturated standard cell. Any difference between these two voltages is detected by an optical galvanometer. Light from the galvanometer mirror falls on a pair of phototransistors that provide the feedback signal to control the load current. The modifications to the current source consisted of changing the reference resistor and changing the bias conditions of the phototransistors. A rotary switch was installed in the circuit. Each switch position inserts a different reference resistor and bias resistor into the circuit.

A commercial relay scanner is used for automatic selection of the appropriate potential terminals of any resistor in the primary circuit of the DCC during a test run. The scanner contains latching-type relays with precious metal contacts for low thermal EMF operation. Shielded two-conductor PTFE-insulated cable is used to make the interconnections between the scanner and the potential terminals of the resistors. The scanner is controlled by the PC through the IEEE-488 bus.

The PC and data handling is similar to the one used in the automated 1 Ω system. The NIST program "AMPERE" on the server computer is executed to compute, analyze, and store the results of the test run. This program adjusts the predicted value of the group mean of the two reference working resistors for variations of the oil bath temperature from the nominal value of 25.000 °C. The corrected group mean is then used to calculate the values of the resistors in the measurement string for the test run.

4.4 Manual DCC Resistance Bridge

The manual type of DCC resistance bridge was used until 1997 for the calibration of Rosa type 1 Ω and 10 Ω standard resistors or their equivalent, and for all four-terminal standard resistors having nominal values of 0.1 Ω, 0.01 Ω, 0.001 Ω, and 0.0001 Ω. The bridge is also capable of measuring 100 Ω and 1000 Ω four-terminal standard resistors. Various ratios of the manual DCC systems have been compared to similar ratios of the automatic DCC system. While the newer DCC bridge is preferred for measurements within its operating range (see section 4.5), manual DCC bridges remain in service at NIST. One such manual DCC bridge system with a high-current range extender is used for calibrations of resistors at currents of 100 A and above.

4.4.1 Description of System

A partial schematic diagram of the manual DCC resistance bridge is shown in Fig. 4.4. The bridge consists of an adjustable 1111.111 turn winding in the primary circuit, and a fixed 1000 turn and adjustable deviation windings in the secondary circuit. The resistor to be measured is connected in the primary circuit and a reference resistor, R_S, is connected in the secondary circuit. If the deviation winding is set equal to the correction c_S of resistor R_S, the bridge becomes direct reading in ohms. The reference resistors and low-power resistors are located in a constant temperature oil bath maintained at (25.00 ± 0.01) °C. The DCC resistance bridge is balanced by adjusting the primary turns ratio for a null condition on detector D using the reversal balancing procedure.

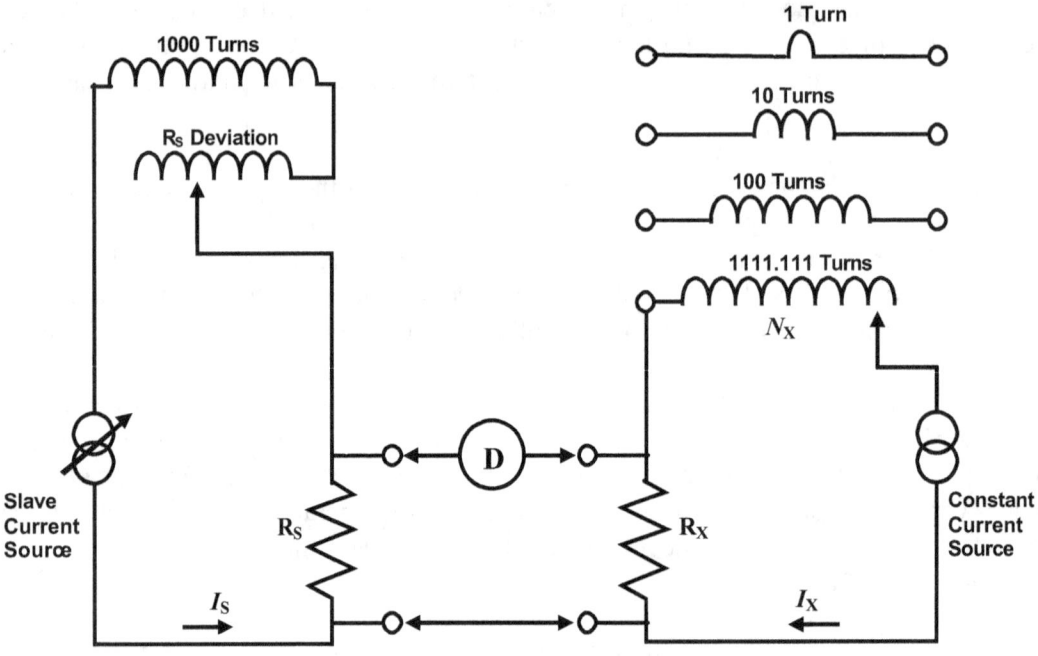

Figure 4.4. DCC resistance bridge with manually selected turns ratio.

The fully adjustable ratio winding in the primary circuit can be replaced by a fixed winding of 100 turns, 10 turns, or 1 turn which will provide additional bridge ratios of 10-to-1, 100-to-1, and 1000-to-1. For these higher ratios, the adjustable fractional-turn section in the primary side is switched to the secondary side in order to balance the bridge. The bridge has a resolution of 0.1 ppm for all ratios; however, it only has a range of 1111.1 ppm for ratios > 1-to-1. For resistance ratio measurements of 1-to-1 and 10-to-1, the adjustable 1 A internal power supply is connected in the primary circuit. When the bridge is used for ratio measurements of 100-to-1 and 1000-to-1, the internal power supply is replaced by an external, adjustable 100 A supply. The 100-to-1 and 1000-to-1 ratios have respective maximum current ratings of 20 A and 100 A.

4.4.2 Measurement of High-Current Standard Resistors

High-current standard resistors (current shunts) are typically designed to operate in air, often at elevated temperature due to significant internal power dissipation (loading). A fixed-ratio 1000-to-1 range extender, two 1000 A power supplies and a high-current pneumatic reversing switch have been installed for measuring these current shunts using a manual DCC bridge. This system is rated to supply up to 2000 A through the current shunt. The high-level current passes through the range extender and a current 1000 times smaller in magnitude is produced using a current null-detector and current source. That current is passed through the fully-adjustable turns winding or a fixed 100 turn winding in the DCC bridge. The reference resistors for this system are operated at much lower current levels (≤ 1 A) and are located in a constant temperature oil bath maintained at $(25.00 \pm 0.01)°C$.

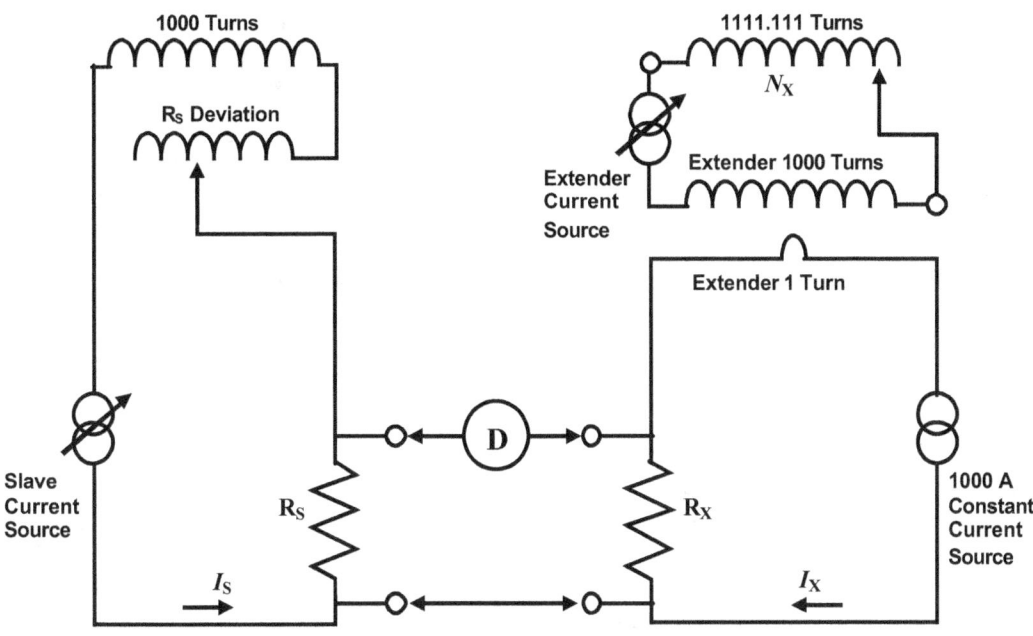

Fig. 4.5. Manual DCC bridge configuration for 100 A to 1000A current in R_X.

The circuit typically used for calibration currents I_X up to 1000 A is illustrated in Fig. 4.5. A bridge current equal to $I_X/1000$ is produced by the range extender bridge output. Adjustable ratio windings with a maximum turns number of $N_X = 1111.111$ can be used to provide a variable DCC ratio, where the direct-reading dial setting $(R_X/R_S)_{dial}$ corresponds to $N_X/1000$. Some examples of this configuration's capabilities are shown in Table 4.1. The maximum current for each ratio is determined by the power dissipation of the reference resistor.

Table 4.1. Manual DCC with 1000-to-1 range extender and 1000 A maximum current.

DCC Ratio	Reference Value, R_S	Nominal R_X (Ω)	$(R_X/R_S)_{dial}$ Nominal	Unknown Value, R_X (Ω)
3000-to-1	$R_S = (1 + c_S) \times 1\ \Omega$	0.000333	0.3333333	$R_X = 0.001 \times R_S \times (R_X/R_S)_{dial}$
2500-to-1	$R_S = (1 + c_S) \times 0.1\ \Omega$	0.000040	0.4000000	$R_X = 0.001 \times R_S \times (R_X/R_S)_{dial}$
1000-to-1	$R_S = (1 + c_S) \times 0.1\ \Omega$	0.000100	1.0000000	$R_X = 0.001 \times R_S \times (R_X/R_S)_{dial}$
1000-to-1	$R_S = (1 + c_S) \times 0.01\ \Omega$	0.000010	1.0000000	$R_X = 0.001 \times R_S \times (R_X/R_S)_{dial}$

At calibration current levels between 1000 A and 2000 A, the output of the range extender would exceed the 1111.111-turn winding maximum measurement current using the bridge as shown in Fig. 4.5. Instead a modified circuit is used to divide the range extender output current by 10, using the 100 turn fixed winding of the DCC, as shown in Fig. 4.6. Variable ratios are obtained by placing these adjustable ratio windings (see Fig. 4.6) in the secondary circuit.

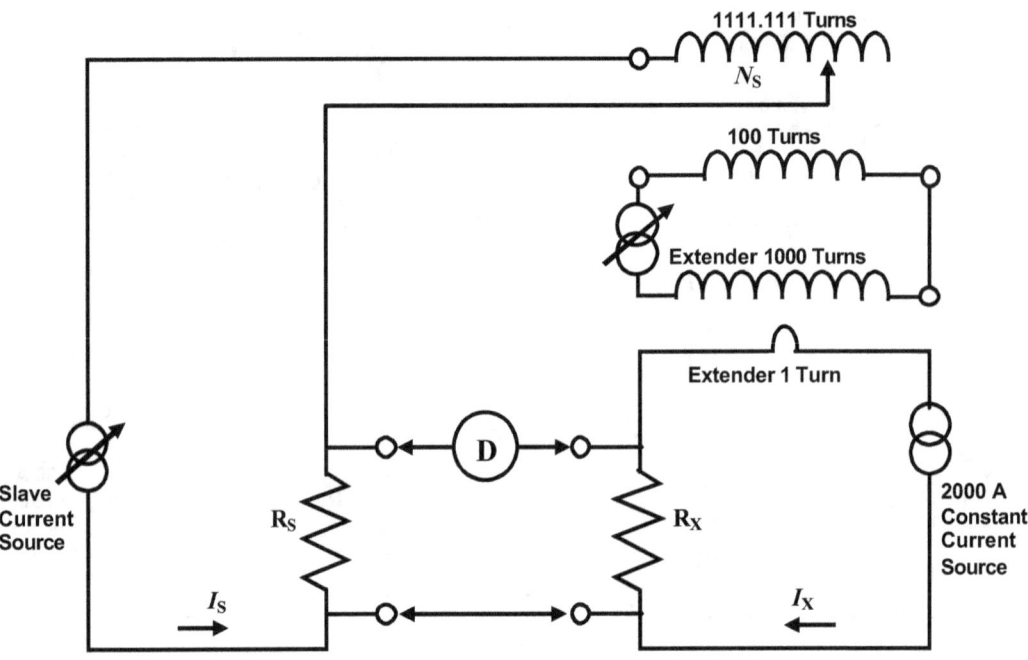

Fig. 4.6. Manual DCC bridge configuration for 1000 A to 2000 A.

Current levels from 100 A to 1000 A can also be used with this special ratio configuration. The resistance ratio calculation (see Table 4.2) is based on a formula similar to Eq. 4.2,

$$R_X = (N_P / N_S) R_S \ .$$

With this circuit configuration, as shown in Fig. 4.6,

$$R_X = (1/1000) \times (100/N_S) R_S \ .$$

It is important to note that the direct-reading dial (of turns N_S) now is used in the R_S side of the bridge. Thus this setting $(R_X/R_S)_{dial}$ corresponds to $N_S/1000$, and as a consequence the calculation *divides* the reference value R_S by the setting of the dials. As shown in Table 4.2, the circuit allows higher ratios, with a maximum of 10000-to-1, and thus reduces the current in the main bridge and the power dissipation in the reference resistor, compared to the circuit of Fig. 4.5.

Table 4.2. Manual DCC with 1000-to-1 range extender connected to $N = 100$ winding.

DCC Ratio	Reference Value, R_S	Nominal R_X (Ω)	$(R_X/R_S)_{dial}$ Nominal	Unknown Value, R_X (Ω)
2500-to-1	$R_S = (1 + c_S) \times 0.1\ \Omega$	0.000040	0.2500000	$R_X = 0.0001 \times R_S / (R_X/R_S)_{dial}$
10000-to-1	$R_S = (1 + c_S) \times 1\ \Omega$	0.000100	1.0000000	$R_X = 0.0001 \times R_S / (R_X/R_S)_{dial}$
5000-to-1	$R_S = (1 + c_S) \times 0.1\ \Omega$	0.000020	0.5000000	$R_X = 0.0001 \times R_S / (R_X/R_S)_{dial}$
10000-to-1	$R_S = (1 + c_S) \times 0.1\ \Omega$	0.000010	1.0000000	$R_X = 0.0001 \times R_S / (R_X/R_S)_{dial}$

4.5 Automated Binary DCC Bridge

Two commercial automatic DCC bridges are used in many calibrations for resistance values up to 1000 Ω. This includes all tests for non-decade value resistors, but excludes comparisons for 1 Ω Thomas-type and certain 100 Ω standards that are measured using the DCC potentiometer bridges, and shunt calibrations made on a manual DCC bridge above the 300 A current level.

The design principles of the automated system are the same as those of the manual DCC bridge, except that direct-reading decade dials for the full and fractional turns are replaced by electronic relays. A turns-selecting relay interface controls the operation of the bridge balance. These relays are used to select from a binary set of real-turn primary windings and fractional-turn primary windings. Fractional turns are achieved by precisely subdividing the primary current into binary parts (1/2, 1/4, 1/8, etc.), with a minimum setting of one part in 16 384 of a real turn.

4.5.1 Measurement Sensitivity

The comparator resolution is limited by the following factors:

1. The voltage noise of the nanovolt detector is 7 nV to 8 nV.

2. The current noise of the comparator is about 40 nA-turns. The secondary winding has 800 turns, giving a typical current sensitivity of about 50 pA.

3. The smallest incremental ratio adjustment is about 0.07 ppm for 1-to-1 ratios and about 0.007 ppm for 10-to-1 ratios. Ratios between 1-to-1 and 10-to-1 provide better resolution and lower uncertainty than ratios below 1-to-1, because the higher ratios use a greater number of turns in the primary winding.

4. The maximum current rating of the secondary ratio windings is about 150 mA. Higher currents in the primary side of the bridge are achieved by using automated range extenders. A 100 A range extender provides a precise 10-to-1, 100-to-1, or 1000-to-1 step-down of current with ranges of 1 A, 10 A, and 100 A, respectively. The new 400 A range extender not only allows higher primary current, but increases the maximum ratio to 1000000-to-1. The maximum current supplied by the range extenders to the primary winding of the DCC unit is 100 mA.

In order to obtain sufficient sensitivity with the manual DCC systems, the power dissipation level can be set at 100 mW for unknown resistors designed to withstand higher currents. Because the automated DCC bridges have higher sensitivity, power dissipation now can be limited to 10 mW for all reference resistors and for unknown resistors, except those used as current shunts.

4.5.2 DCC System Software Description

The commercial software supplied with the automatic bridge allows the user to select and save certain measurement parameters as well as "identifying data" on the standards used. This and other measurement data, including the measured resistance ratio and the date and time of test, are saved in a text file and then transferred to a central database that can be searched and accessed locally by NIST resistance calibration staff. The most significant measurement data parameters are described below.

Resistors R_s and R_x are the standards compared. R_s is the resistor to measure in the secondary (slave) arm of the current comparator, where the number of turns is fixed. Either R_s or R_x may be considered to be the reference standard (see Table 4.3). The measured ratio is determined by the values R_s and R_x of these resistors.

Primary current is the bridge current I_x to be used for calibration of resistor R_x. Since DCC bridges operate with equal voltage across the reference and unknown resistors, the reference resistor R_s must be rated for use at a current of $I_s = I_x R_x / R_s$. This secondary current must not exceed the bridge specification.

Settle time (t_{settle}) is used to control a time delay after every current reversal. This delay is in addition to the bridge reversal, balancing and communications time of up to 12 s. Typically, a settle time of 8 s is used. Longer settle times are used for calibrations of high-current shunts.

Number of measurements (M) controls the total number of resistance ratio measurements in the sequence. Up to 250 measurements may be made over a period of two hours or more for calibrations of high-current standard resistors, so that the effect of internal power dissipation on the temperature of the resistor can reach equilibrium.

Number for statistics (S) controls how many measurements are used to calculate the average ratio. This number is counted backwards from the last measurement in the sequence. The initial five measurements are always discarded to allow time for automatic adjustments to bring the bridge into balance. High-current shunts are measured at thermal equilibrium (see section 9.7).

Table 4.3. Typical automated DCC Bridge system settings

R_x (function)	R_s	I_x	t_{settle}	M	S
1 Ω (reference)	0.1 Ω	10 mA	8 s	30	25
1 Ω (unknown)	1 Ω	100 mA	8 s	30	25
10 Ω (unknown)	1 Ω	10 mA	8 s	30	25
100 Ω (reference)	10 Ω	3.16 mA	8 s	30	25
100 Ω (unknown)	100 Ω	10 mA	8 s	30	25
1000 Ω (unknown)	100 Ω	3.16 mA	8 s	30	25

Table 4.4. Typical automated DCC Bridge system settings with 100 A range extender

R_x (Range extender setting)	R_s	Extender current	t_{settle}	M	S
0.1 Ω (Range 1 A)	1 Ω	0.316 A (10 mW)	8 s	30	25
0.01 Ω (Range 10 A)	1 Ω	1 A (10 mW)	8 s	30	25
0.001 Ω (Range 100 A)	1 Ω	3.16 A (10 mW)	8 s	30	25
0.001 Ω (Range 100 A)	1 Ω	100 A (shunt)	20 s	250	25
0.0001 Ω (Range 100 A)	0.1 Ω	10 A (10 mW)	8 s	30	25
0.000333 Ω (Range 100 A)	0.5 Ω	100 A (shunt)	20 s	250	25

Standard resistors of nominal value 1 Ω or 100 Ω are the working reference standards for the automated DCC bridge system calibrations. A stable intermediate standard may be calibrated using this system, and then used to calibrate an unknown resistor provided that the power dissipation is similar in the two measurements. Intermediate standards are used with the 100 A range extender for calibrating all types of standard resistor with value below 0.0005 Ω and for some non-decade value resistor calibrations. Since a wider range of ratios may be used with the 400 A range extender, its use will allow the working reference standards to be used in essentially all such calibrations.

The measured ratio reported by the DCC software is used to calculate the value of the unknown resistor, based on the known drift factors of the reference resistor. Typically the customer standard is compared to two working standards in succession, and the measurements are repeated on three or more days over a period of one to two weeks, depending on the length and type of the test.

5. RESISTANCE-RATIO BRIDGES

Before the development of the DCC, the comparison of standard resistors was done chiefly by the use of bridges with resistance ratio arms. For many years at NIST, a Wenner bridge [11] served this purpose for the comparison of multiples and submultiples of the ohm from 100 µΩ to 1 MΩ. This bridge, designed in 1918, was a combination Wheatstone bridge and Kelvin double bridge with ratios of 1-to-1 or 10-to-1 having an adjustment range of 5000 ppm. In later years, it was modified to increase its resolution to 0.01 ppm. The bridge was retired in 1983 because of the deterioration of its switch contacts and insulation. By this time it had been already replaced by DCC systems for the comparison of four-terminal standard resistors from 100 µΩ to 1 kΩ as described in section 4.

During the late 1960's and early 1970's, several resistor manufacturers developed a high-quality, transportable 10 kΩ standard resistor in order to establish an additional reference level of resistance along with the traditional 1 Ω level [12]. The 10 kΩ level was attractive because it was nearer to the mid scale of precision resistor measurements, and it was expected to provide a convenient level for comparison with the SI ohm derived from absolute ohm experiments. The improved 10 kΩ standard resistor surpasses the Thomas-type resistor in temperature coefficient of resistance, is about equal to it in transport effects, but is second to it in time stability. Nevertheless, it has gained wide acceptance as a primary reference standard and many of these standards are calibrated at NIST in the dc resistance area. Around 1975, NIST constructed a modified Kelvin-type bridge specifically for the comparison of the improved or "special" 10 kΩ standard resistors [13]. The bridge employed a guard network to reduce errors caused by possible leakage currents in the circuit. In the mid 1990's an automated Warshawsky bridge system [14] was constructed. This automated, guarded measurement system features automatic switching of 10 kΩ resistors using so-called British Post Office (BPO) connectors, and is described in section 5.2. A second guarded Warshawsky bridge system of the same design has been constructed and is used for high-quality air-type resistors that are now available at the 1 MΩ level.

Prior to 1989, standard resistors having nominal values from 1 kΩ to 1 MΩ were calibrated at NIST in a Wheatstone bridge circuit using a manually operated Direct-Reading Ratio Set (DRRS). Generally, standard resistors in this range are of the Rosa design, and are measured in a mineral-oil bath controlled at a temperature of 25.00 °C. Rosa-type standards are now calibrated using an automated system based on an unbalanced-bridge method [15]. Since this method is unique in that it is not a 4-arm bridge method, it will be described separately in section 6.

Standard resistors with nominal decade values from 10 MΩ to 1 TΩ are calibrated at customer-specified voltages and often have characteristics that are not usually observed in lower-valued resistors. These resistors may be wire-wound, at 10 MΩ to 100 MΩ values, or film-type, at these or higher resistance values. Many, especially the film-type standards, exhibit a significant change in resistance with the measurement voltage, which can be as high as 1000 V. A variation of the Wheatstone bridge with two resistance arms replaced by precision voltage calibrators is used in this resistance range. It replaces an earlier NIST teraohmmeter instrument [13] as well as the guarded Wheatstone bridge and is described in section 7. The voltage sources are immune to leakage errors, and are controlled by a computer algorithm to balance the bridge. This active-arm bridge system can be used over a wide range, from below 1 MΩ to above 10 TΩ.

The operation of direct-ratio resistance measurement systems for the comparison of standard resistors are based on two well-known bridge circuits, namely the Wheatstone bridge and the Warshawsky version of the Kelvin double bridge. The Wheatstone bridge circuit can be used for the measurement of two-terminal resistors, while the Warshawsky bridge circuit is used for the comparison of four-terminal resistors. The variations on these bridges presently in use at NIST make possible full automation of resistance measurements.

5.1 Theory of Operation

A schematic diagram of a Wheatstone bridge circuit for the comparison of two-terminal resistors is shown in Fig. 5.1. Resistors A and B constitute the two direct-reading ratio arms. The 1-to-1 resistance ratio of a typical DRRS can be adjusted over a range extending from −5000 ppm to +5000 ppm in increments of 0.1 ppm. R_d is an auxiliary or dummy resistor of the same nominal value as the test or unknown resistor, R_x, and the standard resistor, R_s. Lead resistances l_1, l_2, and l_3 complete the bridge network. Variability in the contact resistances at the resistor terminations are negligible for two-terminal resistor comparisons above 100 Ω. The substitution technique (see section 3) is used for determining the small difference between the resistor values R_x and R_s.

With R_x in the circuit, the bridge equation at balance with a DRRS reading of D_1 in ppm is

$$(A/B) \cdot (1 + D_1 \cdot 10^{-6}) = (R_x + l_1)/(R_d + l_2 + l_3) \ .$$

Similarly for R_s substituted in the circuit, the bridge equation is

$$(A/B) \cdot (1 + D_2 \cdot 10^{-6}) = (R_s + l_1)/(R_d + l_2 + l_3) \ .$$

where D_2 is the new DRRS reading in ppm after the bridge is re-balanced. The value of R_s can be represented as

$$R_s = R'_s\left(1 + c_s \cdot 10^{-6}\right) ,$$

where R'_s is the nominal value of the standard resistor and c_s its correction in ppm. Solving for R_x and neglecting 2nd order and higher terms results in

$$R_x \approx R'_s\left(1 + (D_1 - D_2)\cdot 10^{-6} + c_s \cdot 10^{-6} + (l_1/R_s)\cdot(D_1 - D_2)\cdot 10^{-6}\right) . \qquad (5.1)$$

Usually the last term in the above equation is negligible for measurements above 100 Ω, since the differences between nominally-equal resistors are within 500 ppm, and for most practical circuits the ratio $l_1/R_s < 10^{-3}$. Therefore the Eq. 5.1 for R_x reduces to

$$R_x \approx R'_s\left(1 + (D_1 - D_2)\cdot 10^{-6} + c_s \cdot 10^{-6}\right) .$$

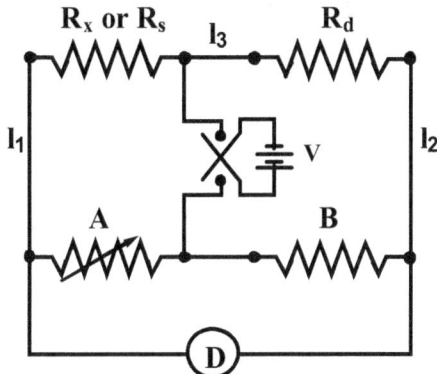

Fig. 5.1. DRRS Wheatstone bridge circuit.

The difference in the DRRS ratio for the two balances gives the difference between the values of the two resistors in ppm. For measurements > 1 MΩ, a guarding network should be added to the Wheatstone bridge circuit in order to suppress errors caused by leakage currents flowing from the main bridge arms to ground potential. The guard circuit drives the shields of the bridge terminations at the proper potentials to confine leakage currents within this network.

The fundamental difference between the Warshawsky bridge and the Wheatstone bridge is the use of auxiliary resistor pairs at each of the four ratio arm junctions to reduce the effect of random junction resistances, which are not included in the above circuit evaluation. These fan resistors help to eliminate the dependence on l_1 observed in Eq. 5.1. The useful range of operation can be extended to lower values of resistance in the ratio arms, and the bridge design can be adapted to measurement systems with higher variation in junction resistance.

5.2 Automatic Guarded Warshawsky Bridge

The first fully-automated NIST resistance bridge of the Warshawsky design was built in the mid 1990's. The bridge used in this four-terminal 10 kΩ resistor measurement system is shown in Fig. 5.2 and 5.3. Fig. 5.2 depicts the system with a feedback circuit used to achieve a null balance at the detector. This active feedback system can be replaced with an optically isolated detector that has sufficient linearity so as not to require a null condition for accurate measurements (Fig. 5.3).

The Warshawsky design's principle advantage (due to the three pairs of fan resistors) is that the resistance in the leads l_1, l_2, and l_3 has very little effect on the measurement results. The fan resistors at l_3 ensure a condition of zero voltage across that potential lead. The fan resistors act as voltage dividers across the two current-carrying leads (l_1 and l_2) and reduce the effect of those leads significantly. It is thus more accurate than the Wheatstone bridge for four-terminal resistance comparisons at intermediate values.

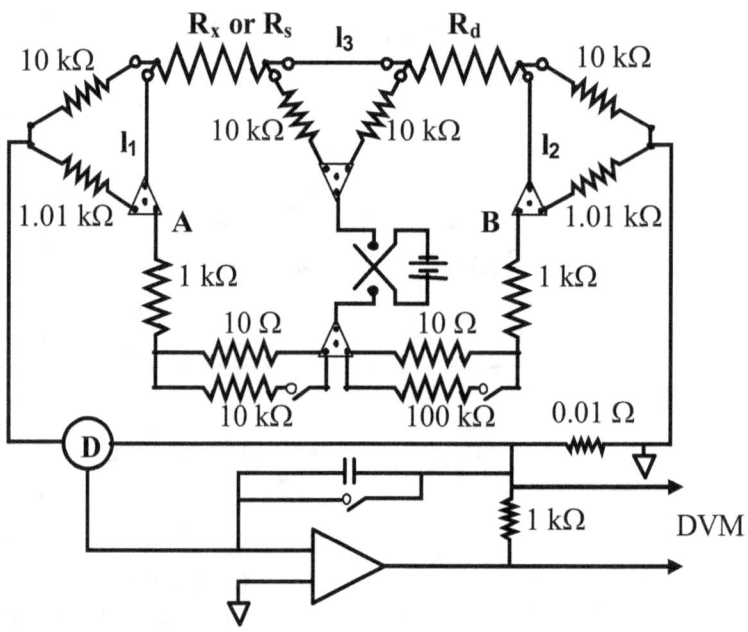

Fig. 5.2. Warshawsky bridge circuit with automatic balance.

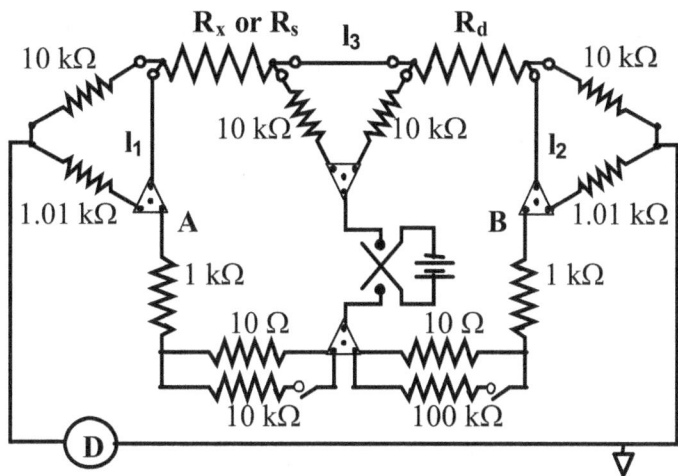

Fig. 5.3. Warshawsky bridge circuit as used with optically isolated detector.

A necessary element of the measurement procedure is to introduce a known offset into the bridge so that the detector readings can be calibrated and ultimately converted into a deviation from the nominal 10 kΩ value for the unknown resistors. The 10 Ω bridge resistors shown in the figures and additional resistors and relays provide approximately 10 ppm or 1 ppm offsets in the bridge balance, and are contained in several aluminum boxes immersed in the 25 °C oil bath. Four-terminal junctions and BPO-type connectors are used to provide access for calibrating fan resistors, A and B arm resistors, and most critically, the 10 Ω resistors with and without the offsets. The active feedback circuit depicted in Fig. 5.2 is contained in another box within the oil bath, and can be removed from the bridge when using the isolated detector.

This bridge circuit for the measurement of special 10 kΩ resistors is guarded (guard circuit not shown in figures). The guard circuit suppresses the effects of leakage currents flowing to ground potential or to other parts of the circuit that may cause an error in the measurement. Unfortunately, the terminations of most resistors under test are not guarded. The shields of the cables connecting a resistor into the test circuit are driven at the guard potential to complete the guard circuit and provide as much suppression of leakage current as possible.

Dummy resistor R_d is an oil-type, 10 kΩ resistor of equal quality as the resistor under test. A temperature lag box houses air-type 10 kΩ resistors (two working standards and one check standard) used in the measurement process. This lag box helps to minimize the effect of fluctuations of the ambient temperature on the measurements. Air-type resistors are calibrated in open air, on a shelf above the lag box. The temperatures of all of the air-type resistors are measured using a digital thermometer with calibrated thermistor probes. The resolution of the temperature measurements is 0.01 °C with uncertainty of ±0.05 °C. An additional check standard is located in the oil bath along with the oil-type resistors under test. The oil bath temperature of (25.00 ± 0.01) °C is monitored continuously by a high-accuracy thermistor thermometer and periodically by a PRT measurement system.

Resistors are measured by the substitution technique, where the working standard and unknown resistors are indirectly compared by substitution into the R_x arm of the bridge circuit. Automated switching of resistors is achieved by a computer-controlled guarded coaxial connector panel [16]. A mechanical positioning system is used to move a 4-connector arm over a panel of 72 connectors mounted in the XY plane. The arm is then lowered in the Z-direction and connects the guarded, low-contact-resistance, BPO plugs to sockets for the appropriate resistor (R_x or R_s). Data are corrected using the known drift factors and temperature coefficients of two working standards (R_s), both of which are compared at the beginning and end of daily measurement runs.

Thermal EMF signals are low compared to some other automated systems in which relays are used to connect the main resistors into the circuit, since the plug-socket connectors have thermal EMF signals below 10 nV. Contact and interconnecting cable resistance values are matched. Each customer resistor is measured over a period of about two weeks, using at least two sets of the 18 possible sets of connectors as a check for possible faults in the connecting circuit.

By modifying the fan resistances and the dummy resistor value, this type of system can be used for comparing other nominally-equal resistors in the range 1 kΩ to 1 MΩ. A second guarded Warshawsky bridge was constructed in 2003 and is used for comparing 1 MΩ standard resistors. The bridge design used at 1 MΩ is shown in Fig. 5.4. This bridge is better suited than the Ring Method (section 6) for comparing air-type resistors, which require external cable connections to the ring stands and thus may introduce additional noise and leakage. Comparisons based on a group of air-type 1 MΩ standards and oil-type Hamon devices have shown that the Warshawsky bridge provides improved uncertainty in all 1 MΩ calibrations. Like the 1 MΩ Ring Method, this system derives the value of its working standards from CCC comparisons based on the QHR standard.

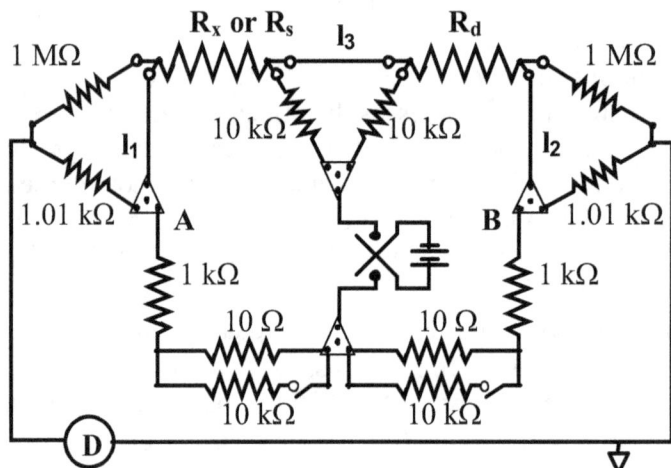

Fig. 5.4. Warshawsky bridge for 1 MΩ comparisons.

6. RING METHOD

This automated measurement system, based on an unbalanced-bridge technique, was developed to replace the manual DRRS system used to calibrate resistors in the resistance range 1 kΩ to 1 MΩ. This technique is referred to as the "Ring" Method since the resistors are connected in a ring configuration. The system is specifically designed to measure the differences among six nominally-equal, four-terminal standard resistors of the Rosa type that are mounted on a mercury ring stand located in a temperature-controlled oil bath. The system has sufficient switching capability to operate four six-arm mercury ring stands, usually at the nominal resistance levels of 1 kΩ, 10 kΩ, 100 kΩ, and 1 MΩ. The system has the flexibility to accommodate other types of standard resistors including those operating in a laboratory air environment. The resistors in the ring are energized at a power level of 10 mW or less.

6.1 Theory of Operation

In general, the unbalanced-bridge or Ring Method consists of connecting an even number of six or more nominally-equal, four-terminal resistors in a ring configuration, i.e., a string of resistors in a closed electrical circuit where each resistor including the first and last resistors in the string are connected at the current terminals. The number of resistors equals $2 \times (n+2)$, where n is an integer ≥ 1. A voltage is applied across opposite corners of the symmetrical ring which divides the ring into two parallel branches, each containing $(n+2)$ resistors. Then, a DVM is used to measure voltages between opposite potential terminals of the resistors that are at nearly equal potentials. Next, the applied voltage points across the ring are rotated in a clockwise (or counterclockwise) direction to the next pair of resistor connection points. Again voltage measurements are taken between corresponding terminals of the resistors that are at nearly equal potentials. This measurement process is repeated $(n+2)$ times. From the $(n+2)$ subsets of voltage measurements, one obtains a set of $3 \times (n+2)$ linear equations that can be solved using a least-squares technique. Values of the resistors can be calculated if the value of at least one of the resistors in the ring is known.

The automated measurement system described in this report is based on a ring of six resistors (n = 1) as shown in Fig. 6.1. Two of the resistors are working standards (R_3 and R_6), one is a check standard (R_4), and the remaining three are additional NIST resistors or resistors under test (R_1, R_2, and R_5). The three pairs of applied voltage points are designated as AA', BB', and CC'. Three subsets of voltage measurements are taken to determine the values of the unknown resistors.

6.1.1 First Subset of Voltage Measurements

The arrangement of the six nominally-equal resistors for the first subset of voltage measurements is shown in Fig. 6.2a. The supply voltage is applied across points A and A'. The voltage differences measured by the DVM across opposite potential terminals of the resistors are designated V_1 through V_6. These voltages are the average of measurements for the two directions of current; this procedure reduces errors caused by thermal EMFs in the critical low-voltage circuits. From this first subset of voltage measurements, one obtains three linear equations of the form analogous to the equations derived for comparing two voltage dividers [17]. These equations, neglecting higher order terms, are:

$$\frac{R_1}{R_2} \approx \frac{R_6}{R_5}\left(1+\frac{V_2-V_1}{V}-\frac{V_4-V_3}{V}\right) ,$$

$$\frac{R_1}{R_3} \approx \frac{R_6}{R_4}\left(1+\frac{V'_2-V'_1}{V}-\frac{V_6-V_5}{V}\right) ,$$

$$\frac{R_2}{R_3} \approx \frac{R_5}{R_4}\left(1+\frac{V'_4-V'_3}{V}-\frac{V'_6-V'_5}{V}\right) ,$$

where V is the nominal voltage across a resistor (approximately 1/3 the applied voltage), and the V'_x's (e.g., V'_1, V'_2, etc.) are repeated measurements of that particular voltage difference to ensure independence. The measurements are made in the sequence implied by the above equations to minimize errors caused by changes in interconnection and lead resistances.

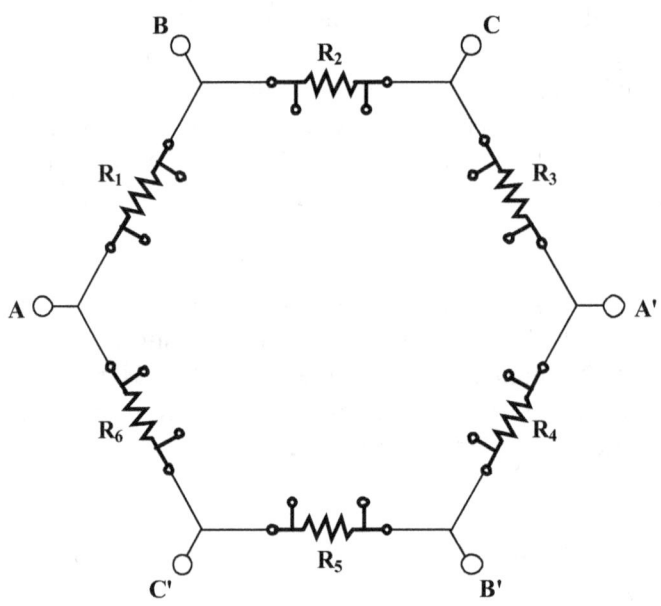

Fig. 6.1. Ring Method circuit for six resistors

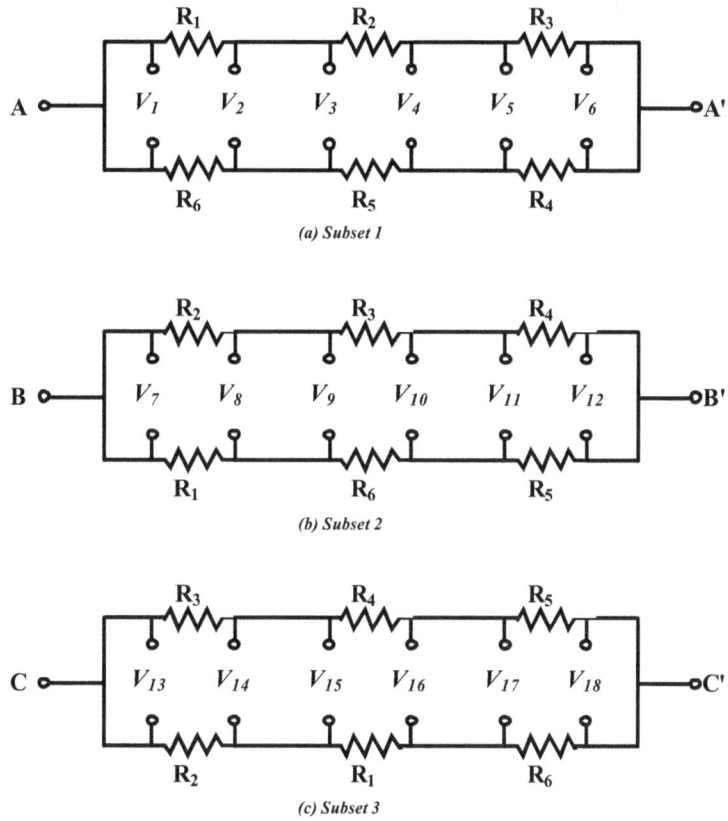

Fig. 6.2. Subsets of voltage measurements for the Ring Method.

6.1.2 Second Subset of Voltage Measurements

The positions of the six resistors, after the supply voltage connections are rotated to points B and B', are shown in Fig. 6.2b. In this figure, the voltage differences measured between opposite terminals of the resistors are designated V_7 through V_{12}. The three linear equations obtained from the second subset of measurements are:

$$\frac{R_2}{R_3} \approx \frac{R_1}{R_6}\left(1+\frac{V_8-V_7}{V}-\frac{V_{10}-V_9}{V}\right) \quad ,$$

$$\frac{R_2}{R_4} \approx \frac{R_1}{R_5}\left(1+\frac{V'_8-V'_7}{V}-\frac{V_{12}-V_{11}}{V}\right) \quad ,$$

$$\frac{R_3}{R_4} \approx \frac{R_6}{R_5}\left(1+\frac{V'_{10}-V'_9}{V}-\frac{V'_{12}-V'_{11}}{V}\right) ,$$

where higher order terms have been neglected.

6.1.3 Third Subset of Voltage Measurements

The positions of the six resistors for the third subset of voltage measurements are shown in Fig. 6.2c. The supply voltage connections are rotated to points C and C'. The voltage differences measured between opposite resistor terminals are designated V_{13} through V_{18}. The three linear equations, neglecting higher order terms, obtained from these voltage measurements are:

$$\frac{R_3}{R_4} \approx \frac{R_2}{R_1}\left(1+\frac{V_{14}-V_{13}}{V}-\frac{V_{16}-V_{15}}{V}\right),$$

$$\frac{R_3}{R_5} \approx \frac{R_2}{R_6}\left(1+\frac{V'_{14}-V'_{13}}{V}-\frac{V_{18}-V_{17}}{V}\right),$$

$$\frac{R_4}{R_5} \approx \frac{R_1}{R_6}\left(1+\frac{V'_{16}-V'_{15}}{V}-\frac{V'_{18}-V'_{17}}{V}\right).$$

6.2 Data Analysis

From the three subsets of measurements, one obtains a set of nine linear equations. These equations can be expressed in terms of the resistor corrections using the relationship,

$$R_x = R'(1+c_x),$$

where x is an integer from 1 to 6 denoting the resistor number, R_x is the value of the resistor, R' is its nominal value, and c_x is its correction. The nine linear equations can now be written as:

$$+c_1 - c_2 + c_5 - c_6 \cong [(V_2 - V_1)/V] - [(V_4 - V_3)/V] = d_1,$$

$$-c_1 + c_3 - c_4 + c_6 \cong [(V'_1 - V'_2)/V] - [(V_5 - V_6)/V] = d_2,$$

$$+c_2 - c_3 + c_4 - c_5 \cong [(V'_4 - V'_3)/V] - [(V'_6 - V'_5)/V] = d_3,$$

$$-c_1 + c_2 - c_3 + c_6 \cong [(V_8 - V_7)/V] - [(V_{10} - V_9)/V] = d_4,$$

$$+c_1 - c_2 + c_4 - c_5 \cong [(V'_7 - V'_8)/V] - [(V_{11} - V_{12})/V] = d_5,$$

$$+c_3 - c_4 + c_5 - c_6 \cong [(V'_{10} - V'_9)/V] - [(V'_{12} - V'_{11})/V] = d_6,$$

$$+c_1 - c_2 + c_3 - c_4 \cong [(V_{14} - V_{13})/V] - [(V_{16} - V_{15})/V] = d_7,$$

$$+ c_2 - c_3 + c_5 - c_6 \cong [(V'_{13} - V'_{14})/V] - [(V_{17} - V_{18})/V] = d_8 \quad,$$

$$- c_1 + c_4 - c_5 + c_6 \cong [(V'_{16} - V'_{15})/V] - [(V'_{18} - V'_{17})/V] = d_9 \quad.$$

where the $d's$ are calculated from the voltage measurements. These equations can be represented in matrix notation as:

$$\begin{vmatrix} 1 & -1 & 0 & 0 & 1 & -1 \\ -1 & 0 & 1 & -1 & 0 & 1 \\ 0 & 1 & -1 & 1 & -1 & 0 \\ -1 & 1 & -1 & 0 & 0 & 1 \\ 1 & -1 & 0 & 1 & -1 & 0 \\ 0 & 0 & 1 & -1 & 1 & -1 \\ 1 & -1 & 1 & -1 & 0 & 0 \\ 0 & 1 & -1 & 0 & 1 & -1 \\ -1 & 0 & 0 & 1 & -1 & 1 \end{vmatrix} \begin{vmatrix} c_1 \\ c_2 \\ c_3 \\ c_4 \\ c_5 \\ c_6 \end{vmatrix} = \begin{vmatrix} d_1 \\ d_2 \\ d_3 \\ d_4 \\ d_5 \\ d_6 \\ d_7 \\ d_8 \\ d_9 \end{vmatrix}$$

The restraint vector is given by:

$$c_s = \begin{vmatrix} 0 & 0 & 1 & 0 & 0 & 1 \end{vmatrix},$$

where c_s is the summation of the corrections for the two standard resistors R_3 and R_6. The solution to obtain the corrections to the other resistors can be calculated using a linear least-squares analysis routine [18].

Each complete data cycle results in resistance values for all six resistors, expressed as the correction from the nominal resistance, and based on the drift factors of two working standards. A measurement run consists of four complete data cycles. The four results are combined by averaging, and a standard deviation is calculated. This and other significant data are stored in the central database. The data histories stored on the database server can be viewed by NIST resistance staff using a web browser or by downloading the data to a spreadsheet program.

6.3 Guard Network

This measurement system's voltage source and DVM can have leakage resistances to ground on the order of 10^{10} Ω to 10^{11} Ω. Leakage resistance at the DVM inputs effectively shunts parts of the ring circuit and could result in significant measurement errors at the 10 kΩ level and above. To reduce these errors, an active guard network is used to drive the guard terminal of the DVM at nearly the same potential as its input terminals. The guard network consists of three resistors in series that are connected across the output of the voltage source as shown in Fig. 6.3. This guard circuit is effective at reducing leakage currents when the DVM is connected to measure V_2, V_3, V_4, or V_5. Guarding the DVM during V_1 or V_6 measurements is not necessary since any leakage current is supplied directly from the source outputs.

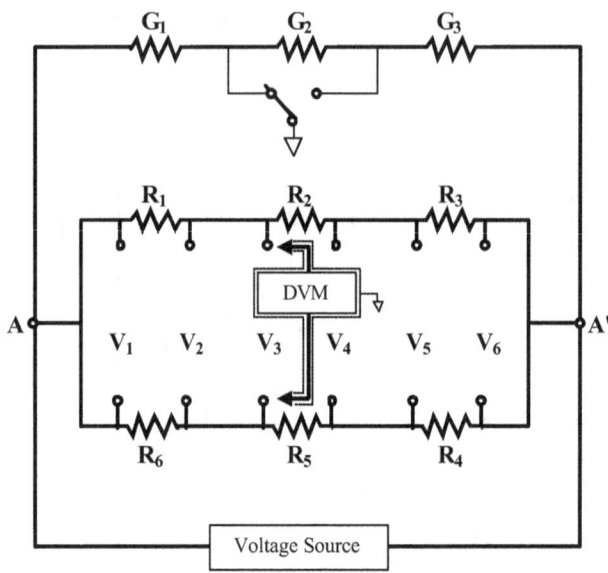

Fig. 6.3. Ring Method with guard circuit.

7. HIGH RESISTANCE MEASUREMENTS

7.1 High-Resistance Standards

The highest value commercially available wire-wound resistor is a 100 MΩ unit. Wire-wound resistors are often packaged as sealed units to protect the elements from the effects of oxygen and moisture in the atmosphere. Generally film-type resistors are not as stable as wire-wound resistors, and usually are provided with an insulator coating but are not hermetically sealed. If the film is open to the atmosphere (not hermetically sealed), moisture can produce two reversible effects: resistance decrease due to surface leakage across the element, and stress-induced resistance change due to swelling in insulators in contact with the resistance film. Higher-resistance standards fabricated at NIST [19] incorporate film-type resistors hermetically sealed in specially-designed containers and heat-treated to improve stability and voltage coefficients.

Many Hamon transfer standards [20,21] have been constructed using groups of ten hermetically sealed high-value resistance units. The wire-wound or film-type resistors used in constructing the transfer standards are reasonably well matched in value, to improve the transfer accuracy of the series and parallel networks. Resistors were selected for the guard networks that are closely matched to the main resistance values. The nominal value of these guard resistors is a factor of 100 smaller than the main resistance elements in most NIST transfer standards above 100 MΩ.

A process of sealing the resistance elements in metal-insulator-metal containers was developed for resistance elements of 100 GΩ and above. These containers allow the metal end-sections to be driven at separate guard potentials nominally equal to the potentials at the resistor terminations, which are insulated by glass-to-metal seals. This guarding method suppresses leakage currents flowing across the glass insulators, which have leakage resistance of order 100 TΩ.

7.2 Guarded Active-Arm Bridge

The active-arm bridge [22] is formed by substituting two of the resistive arms of a Wheatstone bridge circuit with low-impedance digitally-programmable voltage calibrator sources. Standard resistors over the entire range of values from 10 MΩ to above 1 TΩ are measured using a NIST-built guarded active-arm bridge. A manual guarded Wheatstone bridge and a semi-automated commercial teraohmmeter were used for calibrations in this resistance range until 2003, when comparisons between these systems and the active-arm bridge were completed.

For a Wheatstone bridge, shown in Fig. 5.1, the equation at time of balance is

$$A/B = R_x/R_d ,$$

where A and B are the values of resistance in the A and B ratio arms of the bridge and R_x and R_d are the values of the unknown and dummy resistors. The detector D and voltage source V complete the traditional Wheatstone bridge. In the automated active-arm bridge, resistances A and B are replaced with voltages E_1 and E_2 respectively yielding the equation

$$E_1/E_2 = R_x/R_d , \qquad (7.1)$$

at time of balance. A diagram of the NIST guarded active-arm bridge [23] is shown in Fig. 7.1. The bridge voltage is generated by two independent programmable voltage sources, and is not limited by the maximum power dissipation in the ratio arms A and B as it is in the Wheatstone bridge. The bridge voltage ratio together with the small current through detector D is used to calculate the resistance ratio. Only the voltage ratio is critical to this measurement, and this is calibrated at regular intervals using an accurate voltage divider.

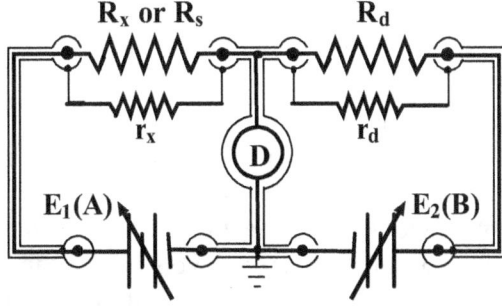

Fig. 7.1. Active-arm bridge with guard resistors r_x and r_d connected to the shields.

The detector D is an electrometer having a high input impedance of $>10^{16}$ Ω. The detector is programmable and has a resolution of ± 3 fA on its most sensitive range. Its low input is connected to the bridge terminal between the voltage ratio arms A and B at the low output of each source. The main bridge arms, drawn with heavy lines, consist of the voltage ratio arms A and B, the dummy resistor R_d, and the test resistor R_x or R_s. The current-reversal method is used to eliminate thermal voltage signals. Drawn with lighter lines and in parallel with the main bridge circuit is an auxiliary guard network for the purpose of reducing measurement errors caused by leakage currents. The guard circuit is driven by connections to the main bridge at the high outputs of the voltage calibrators.

The bridge guard circuit is connected to earth ground at a single point, between the two voltage calibrators in arms A and B. The grounded guard includes the shield of the detector D and the guard circuit at the high and low sides of the detector. NIST-built transfer devices or series-parallel buildup standards typically contain internal guard networks, consisting of ten guard resistors between the shields of the connections for the internal resistance elements. These transfer device guard networks thus take greater advantage of the guard circuit by guarding each of the internal connections.

The test resistor R_x or R_s is measured with the case grounded, so that a leakage resistance path between its high terminal and case is effectively in parallel with the voltage source A, and does not contribute to any measurement errors. Terminal leakage resistance at the low test resistor terminal is effectively in parallel with the detector, and therefore a very small voltage is present to drive leakage. The test resistor is compared to a working standard in the same arm of the bridge (substitution technique), and consequently, any leakage errors associated with R_d are negligible if the leakage resistances are constant during the time interval of the measurement run.

Decade-value standard resistors of nominal value up to and including the 10^{12} Ω level are compared by the substitution method with NIST working standards of the same nominal value. The guarded active-arm bridge also has the advantage of being able to calibrate standards at 10 TΩ and 100 TΩ, two resistance ranges that have not been supported by NIST calibration services in recent years.

Above 10^{12} Ω, the measurements are described as special tests because calibrated bridge ratios of 10-to-1, 100-to-1, and 1000-to-1 are used together with working standards of lower decade value to determine the resistance. In the ratio configuration, a NIST transfer device or working standard of lower value is used in place of the dummy resistor R_d.

The resistors under test and the dummy resistor R_d are placed in an environmental air chamber with temperature and relative humidity control of (23.0 ± 0.1) °C and (35 ± 5) %, respectively. The temperature and humidity are continuously monitored by a digital thermometer and digital humidity transducer. Electrical access into the air chamber is made via 32 BPO connectors mounted on a PTFE panel. Two or three different levels of resistance standards can be connected since this provides for the connection of 16 resistors within the chamber. Usually two working standards, one check standard, one dummy resistor, and one or more customer resistors are connected for each resistance level to be measured.

7.3 Measurement Process

The resistance standards are allowed to equilibrate for at least 24 hours before active-arm bridge measurements. A calibration measurement consists of four cycles of a computer-controlled process comparing working, check, and customer standards at a given resistance level against a dummy standard at the same or lower level. The resistance standards in the A side of the bridge are measured at the customer-specified voltage level E_1. Typically one of the two working standards is measured first during each cycle. The check standard and customer standard(s) are then measured, followed by the remaining working standard. Drift in the system and/or dummy resistor are eliminated by repeating this cycle four times. Four measurement values are calculated for each standard based on values calculated from the drift factors and voltage coefficients of the two working standards. Standard deviations also are calculated from these four sets of results.

7.3.1 Balancing Algorithm

The electrometer senses the difference in the currents, ΔI, flowing through R_x and R_d. Initially, the voltage sources are set to E_1 and E_{2Est}. Rough pre-balance measurements on the unknown standards are used so that E_1 and E_{2Est} give a moderate offset from the nominal ratio of R_x and R_d. A new estimated output of source B which should drive ΔI close to a null is calculated using the equation,

$$E'_{2Est} = \left(-\Delta I + E_1 / R_d\right) R_x \quad . \tag{7.2}$$

The source B is then set to a new value E'_{2Est} after which the current $\Delta I'$ is measured, and as long as $\Delta I' ? \Delta I$, a linear equation can be applied,

$$E_2 = \left(\Delta I \cdot E'_{2Est} - \Delta I' \cdot E_{2Est}\right) / \left(\Delta I - \Delta I'\right) \quad . \tag{7.3}$$

It was found that a positive offset in the initial setting E_{2Est} avoids the possibility of the two measurement currents being nearly equal, and furthermore that an equal-magnitude negative offset in the value of E'_{2Est} makes the algorithm more robust. Once a bridge null has been determined, each of the two measurements are repeated with alternating polarities of the voltage sources, thereby reversing the flow of current and eliminating the effect of constant thermal EMF signals in the detector circuit. The reversal sequence is repeated in a manner that helps to eliminate settling time asymmetry. The unknown value R_x can then be accurately calculated from E_1, E_2 and the similar results for known working standard resistors.

7.3.2 Automatic Resistor Selection

Accurate and fully automated substitution measurements can be performed using a guarded scanner for nominal values up to 10 GΩ. The low-thermal guarded scanner [24] was developed commercially in cooperation with NIST and is used for high-resistance levels up to 1 GΩ. Leakage resistance paths through the relay components or circuit boards are broken by guard conductors or air gaps. These interrupt the leakage paths; for example, the rocker plate actuator of each relay is connected to the guard circuit to isolate circuitry within the relay. Two relays are used for each test resistor, one for the high side and one for the low side measurement circuit. The guard circuit is switched along with the measurement circuit connection using the same relay.

For higher resistance values, a computer-controlled coaxial connector panel using a two-axis positioning system has been developed, based on the guarded coaxial connector panel design used in the Warshawsky bridge. This connects the resistors to both the measurement circuit and guard circuit. The circuits are carried through coaxial cables with PTFE insulator surrounding the inner and outer conductors. For high-voltage applications, the coaxial leads are terminated to triaxial connectors, adding an outer shield that is connected to earth ground.

8. THE U.S. REPRESENTATION OF THE OHM

8.1 Quantized Hall Resistance

The U.S. representation of the ohm (Ω_{NIST}) is based on the quantized Hall resistance (QHR) standard. Before the introduction of the QHR standard on Jan 1, 1990, Ω_{NIST} was maintained using five Thomas-type reference resistors at the 1 Ω level and most transfer of resistance between different levels relied on Hamon device transfer standards. Resistance values in the intermediate resistance range now are transferred using cryogenic current comparator (CCC) bridges. This scaling method can be used to directly compare the QHR standard against standard resistors in one step, which is similar in concept to a DCC ratio measurement. At NIST, reference standards at the 100 Ω and 1 MΩ levels are used for these direct comparisons.

In the early 1990's, a group of five well-characterized Evanohm wire-wound reference standards at the 100 Ω level were selected and these resistors formed the NIST 100 Ω reference group. These five 100 Ω standards are the basis for direct scaling between the 1 Ω and 10 kΩ decade levels. The five resistors in the 100 Ω reference group are periodically compared with the NIST QHR standard, approximately twice a year, and these resistors are concurrently compared against the 1 Ω and 10 kΩ working standards at NIST. Recently, a secondary starting point in decade scaling at the 1MΩ resistance level has been introduced. We now compare 1 MΩ standards directly with the QHR at the 12 906.4035 Ω plateau using a CCC two-terminal bridge method. A diagram of the entire scaling process in use at NIST is shown in Fig. 8.1. On the right-hand side of the figure, NIST calibration test numbers are shown in parentheses.

The QHR device is characterized based on international guidelines when it is used in resistance comparisons [25]. Reference [26] describes some of the processes in use at NIST for CCC measurements and the criteria for selecting QHR devices for use as a QHR standard. CCC scaling to the 100 Ω reference group involves comparing each resistor to the QHR at the 12 906.4035 Ω plateau many times over a two-week or longer period. Primary scaling using QHR comparisons to a group of 1 MΩ reference standards also occurs within this time period.

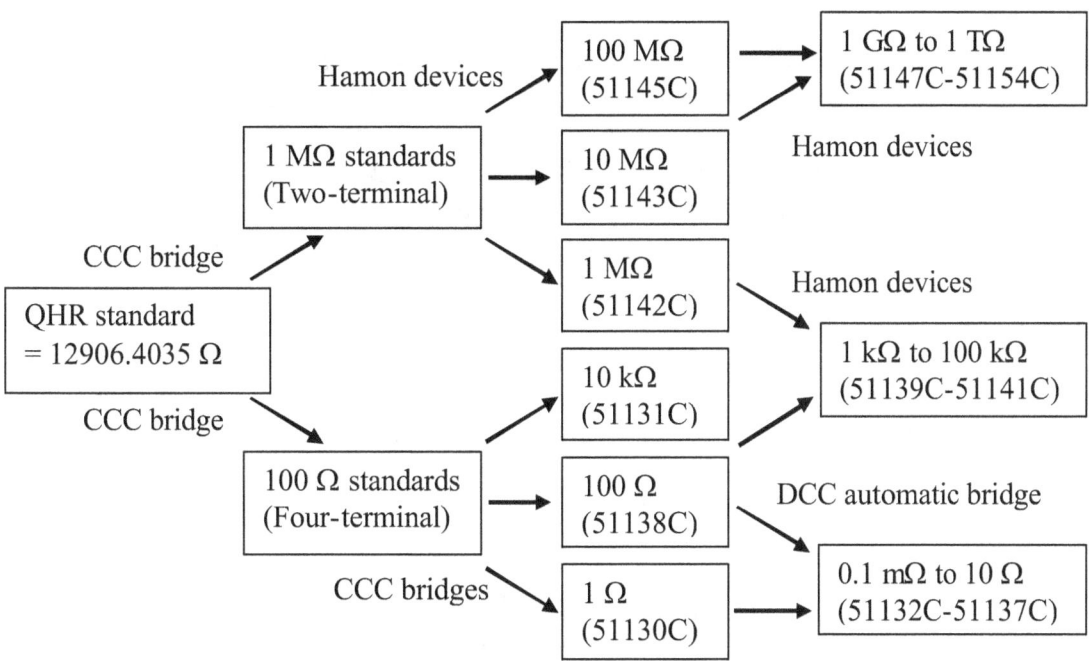

Fig. 8.1. Resistance scaling from the QHR to working standards

8.2 Primary Scaling

Prior to 1992, all scaling to decade-value resistors at the 1 Ω level and above was performed with Hamon devices [27]. Now, assigning reference values to the standards that make up the 10 kΩ, 100 Ω and 1 Ω working groups is performed with CCC bridges, by a method similar to that of the Measurement Assurance Program (MAP) offered at NIST. In particular, this method utilizes the normal 1-to-1 substitution measurement processes. For example, in calibrating the Thomas-type 1 Ω resistance standards, two stable 1 Ω resistance standards with low load coefficients are compared to the 100 Ω reference standard bank using a 100-to-1 CCC bridge, allowing values to be assigned to the 1 Ω standards in terms of the QHR. The 1 Ω DCC potentiometer system calibrates these two standards at night based on the 1Ω working group, concurrent with the daily 100-to-1 scaling measurements. This process is repeated over a period of several days or more.

A linear model of the drift of the 1 Ω working group is calculated from the results of similar periodic scaling measurements spanning two or more years. The drift in value of NIST 1 Ω working standards is in general highly predictable using this linear model, based on prior observations and careful maintenance of the resistors and their environment. For the 1 Ω working group, and generally for other working standards at NIST, a linear equation fitting the scaling data is used to calculate predicted reference values. The linear equation and its predicted values are confirmed or adjusted (as necessary) using each new result from periodic scaling.

The intercept and first-order coefficient of this linear equation define the drift factors of the 1 Ω working group. These constants are adjusted if the predicted values and the values in terms of the QHR differ by more than about 0.015 ppm, as described in section 9.2.

10 kΩ working standards used in the Warshawsky automatic bridge and 100 Ω working standards used with the DCC potentiometer likewise derive their drift factors using the CCC and a MAP method transfer. Two or more 10 kΩ standards are compared to the 100 Ω reference group using a 100-to-1 CCC system over several days. These 10 kΩ standards are also compared to two stable standards used in the 100 Ω DCC potentiometer, so that values based on the QHR are assigned to at least two 10 kΩ standards and two 100 Ω standards on the same day.

The 10 kΩ and 100 Ω automated calibration systems compare these standards to the working standards at night, and use the drift factors of the working standards to assign calibration values. The drift factors of the working groups at 10 kΩ and/or 100 Ω are adjusted if these assigned values differ by more than 0.025 ppm from the values based on the QHR.

8.3 CCC Ratios

The NIST CCC devices are of the overlapped-tube type [28] with a commercial SQUID (superconducting quantum interference device) sensor to detect the ampere-turn condition of the comparator. Due to the Meissner effect the superconducting overlapped-tube is a perfect shield for the magnetic flux produced by small currents in the windings. This eliminates any dependence on the position of the windings, while the SQUID sensor and absence of a magnetic core greatly reduce noise compared to DCC systems.

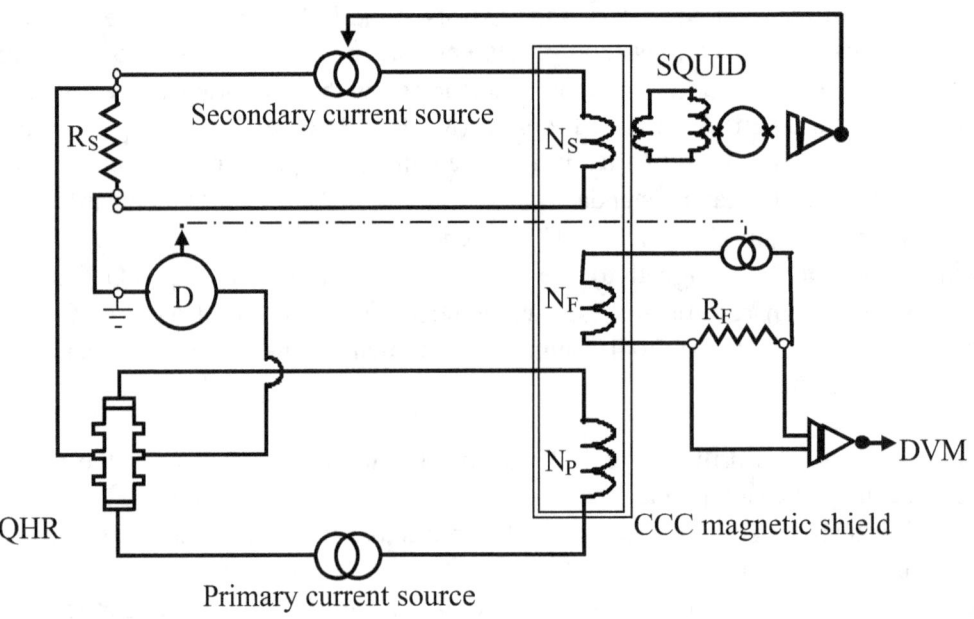

Fig. 8.2. 129.06-to-1 winding-ratio CCC bridge for QHR-to-100 Ω measurements

In a four-terminal CCC bridge (Fig. 8.2), a commercial nanovolt detector, D, senses the voltage difference across the resistors, and provides a feedback current through R_F and N_F. The voltage drop across R_F is monitored with an optically isolated digital voltmeter and is a measure of the difference of the resistor corrections. In order to eliminate leakage currents, the primary and secondary current sources have a common ground at one point and otherwise are optically isolated from one another [29]. The combined relative uncertainty of the 100-to-1 and 129.06-to-1 winding-ratio CCC bridges constructed at NIST is less than 1×10^{-8} for typical measurement conditions [30]. The resistors that comprise the 100 Ω reference group have small temperature coefficients (TCRs) with different signs [31]. Internal power dissipation or "loading" between zero and 10 mW has a negligible effect for the group average resistance. These resistors can be calibrated against the QHR at a variety of different current levels, from about 2 mA up to about 8 mA.

Current comparator bridges apply equal voltage to the resistance standards, such that the standard of lower resistance dissipates higher power internally. This has negligible effect on scaling between the 100 Ω reference group and the 1 Ω working group because the load in scaling is the same as that used in the DCC potentiometer. The 10 kΩ standards are used at a different loading level in scaling. Uncertainty due to the effect of loading can be determined by comparison with Hamon device scaling at equal loading, and by using intermediate 10 kΩ standards with positive and negative TCRs.

The two-terminal CCC bridge used for 1 MΩ scaling makes use of a unique property of the quantum Hall effect. Six contacts provide distinct paths for the measurement current entering or leaving the QHR device [32]. Surprisingly, the multiple connections can result in a two-terminal resistor of value 12906.4035 Ω that does not depend significantly on the resistance of the leads, when the QHR device is connected to the measurement circuit as in Fig. 8.3.

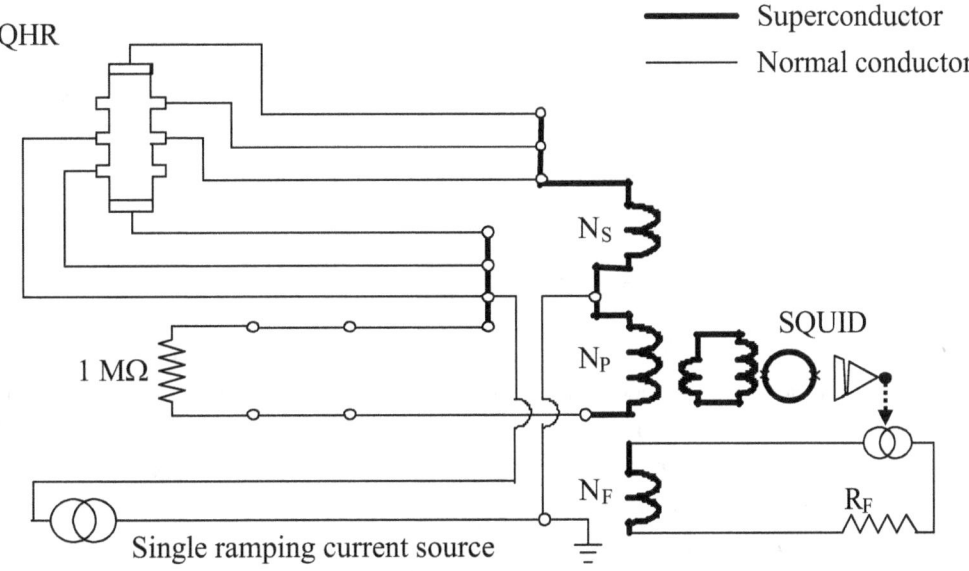

Fig. 8.3. Schematic diagram of two-terminal CCC bridge

Physics related to the direction of the magnetic flux through the device determines exactly how much measurement current flows through each contact. With the correct magnitude and direction of magnetic flux through the QHR device, the current in the set of leads nearest the middle of the QHR device shown in Fig. 8.3 is less than 1 part in 10^6 of the total device current. Leads connected to the second set of QHR terminals, placed diagonally across the device, carry about 0.1% of the total current, with most current flowing through leads connected at the top and bottom. The windings and most of the connections within the two-terminal CCC bridge are superconducting, and do not affect the bridge ratio. However, the lead resistance in the bridge connections to the 1 MΩ standard must be measured separately, and subtracted from the measured resistance on the primary side of the bridge.

8.4 DCC ratios

The high ratio accuracies of the room-temperature DCC result from the use of multiple-winding, high-permeability toroidal transformers. Like the CCC ratio, a DCC ratio is insensitive to lead resistances and current level. The drift factors used to predict the values of the two 1 Ω and two 100 Ω working standards are determined using the DCC potentiometer systems and CCC bridges and are maintained with close agreement to QHR scaling. Direct comparisons using the DCC bridge can check the DCC 1-to-1 ratio using both of the 1Ω or both of the 100 Ω working standards.

The ratio of 10-to-1 is the most significant DCC automatic bridge ratio other than 1-to-1, and is checked by using an intermediate 10 Ω transfer standard in two scaling steps. These 10-to-1 steps are combined and compared to CCC 100-to-1 scaling based on the values of the 1 Ω and 100 Ω working standards.

The range extender ratios are checked by a build-up process. First without using the range extender, the 0.1 Ω standards are compared against 1 Ω working standards in the secondary DCC bridge arm. Then, the 10-to-1 extender ratio is used to scale between 1 Ω, 0.1 Ω, and 0.01 Ω levels. The 100-to-1 ratio is verified using two combined 10-to-1 steps, and the 100-to-1 ratio is used to compare these 0.1 Ω and 0.01 Ω standards to 1 mΩ and 0.1 mΩ standards. Finally, the 1000-to-1 ratio is checked by using it to compare 1 Ω and 0.1 Ω standards to the same 1 mΩ and 0.1 mΩ standards.

8.5 Transfer Standards

NIST transfer standards provide accurate ratios of 10-to-1 and 100-to-1 for extending the use of Ω_{NIST} to higher resistance levels. Special 10-to-1 Hamon devices are used to scale up from 10 kΩ to 100 kΩ and down from 1 MΩ to 100 kΩ, while 100-to-1 Hamon devices are used above the 1 MΩ level. The advantage of transfer standards is that they are calibrated at one resistance level and are then used with equal accuracy at a different resistance level as a short-term working standard. In a series of measurements using transfer standards, the Ω_{NIST} can be extended in multiple decade values up to 10^{12} Ω. Neither the absolute accuracy nor long-term stability of the resistors in the transfer standards has any effect on the measurement accuracy. The concerns with the transfer standards are only with the short-term stability required to complete the measurements and with thermoelectric and connection effects. The transfer accuracy is ensured through proper design of the connections and terminations and various operational tests.

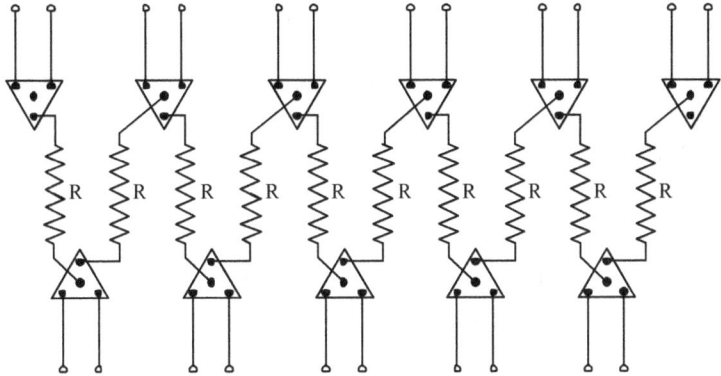

Fig. 8.4. Diagram of a ten-section transfer standard

Figure 8.4 is a schematic diagram of a typical NIST transfer standard. It consists of 10 nominally-equal resistors connected permanently in series by means of "tetrahedral" junctions. Each junction has two current and two potential terminations, and the 4-terminal resistance of each junction is designed to be zero. If necessary, junctions may be adjusted to have a negligible effect on the transfer accuracy. The degree of adjustment depends on the nominal value of the main resistors. Figure 8.5 shows the transfer standard connected in a parallel mode using special

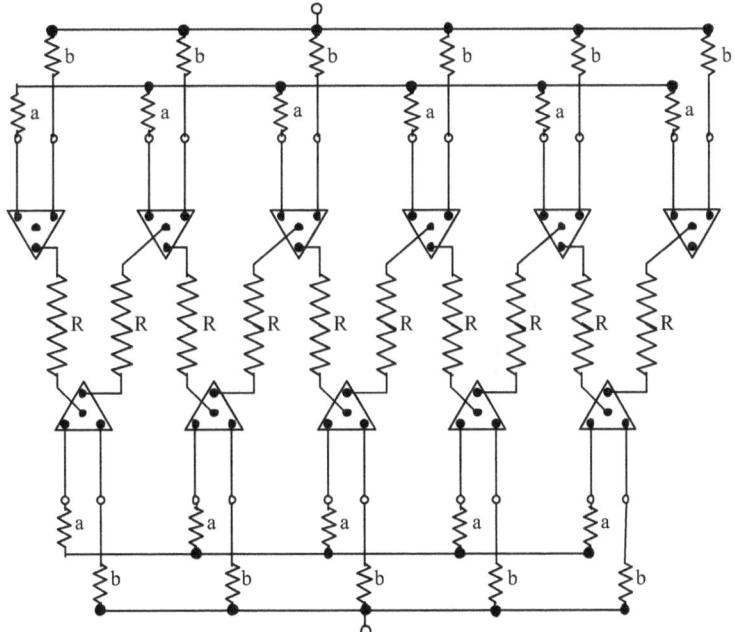

Fig. 8.5. Parallel-mode transfer standard with resistance value $R/10$.

fixtures. The potential circuit of the fixtures contains "fan" resistors for suppressing connection errors. The series and parallel resistances of a transfer standard with nearly equal-valued resistance elements are:

$$R_s = N \cdot R(1+c) \quad ,$$

and

$$R_p = (R/N)\left(1 + c + f(c^2)\right) \quad ,$$

where R_s and R_p are the series and parallel resistances, respectively, N equals the number of resistors, R is the nominal resistance of a single element, c is the correction to the nominal value for the series mode, and $f(c^2)$ includes negligible higher-order terms. For a transfer standard having 10 main resistors,

$$R_s / R_p = N^2 = 100 \quad .$$

Figure 8.6 shows the transfer standard connected in a series-parallel mode for a 10-to-1 ratio. In this configuration only nine of the ten resistors are used. The series-parallel resistance R_{sp} is

$$R_{sp} = R\left(1 + c + 0.1(c_{sp} - c_x)\right) \quad ,$$

where c_{sp} is the correction of the nine resistors in the series-parallel configuration and c_x is the correction of the tenth resistor. The resistance ratio is

$$R_{sp} / R_p = 10\left(1 + 0.1(c_{sp} - c_x)\right) \quad .$$

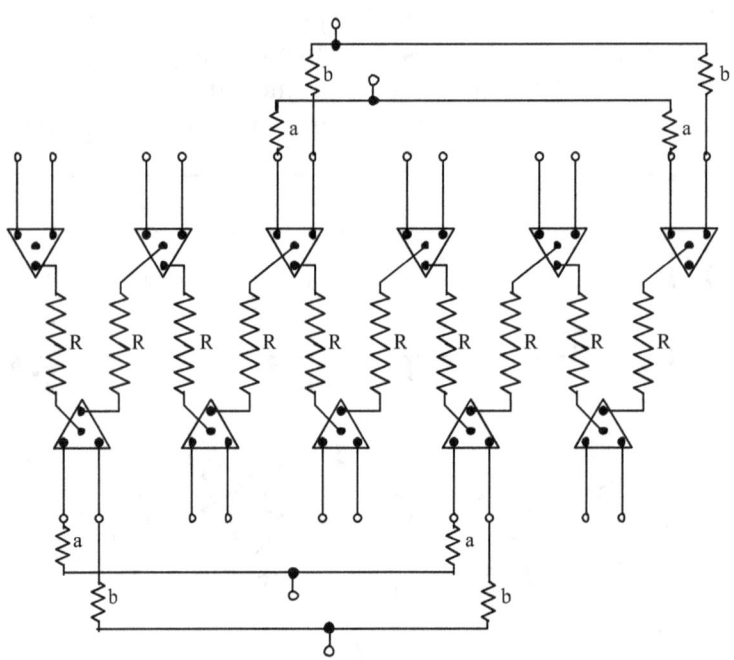

Fig. 8.6. Transfer standard in the series-parallel mode, with resistance value R.

The individual resistance corrections (such as c_x) of the resistance elements are in fact not equal and introduce second-order corrections to the above relationships that are usually negligible. The ten elements can be measured individually as a check on the construction, and the sum of these resistance values should equal the series resistance. In normal practice, only one measurement of the transfer standard is required in the series and parallel modes, while two measurements must be made in the series-parallel mode to completely define the transfer resistance.

Figure 8.7 is a block diagram of the NIST scaling process using Hamon transfer standards, where boxes represent NIST working standards. The "P", "S", and "SP" notation indicates whether the Hamon transfer standard is in the parallel, series, or series-parallel mode, respectively. Arrows indicate the reconfiguration modes which allow the comparisons to be made, and point in the direction that transfer based on the QHR normally occurs using the most accurate scaling techniques. Scaling to the $10^4\ \Omega$ and $10^6\ \Omega$ levels is based on CCC bridges.

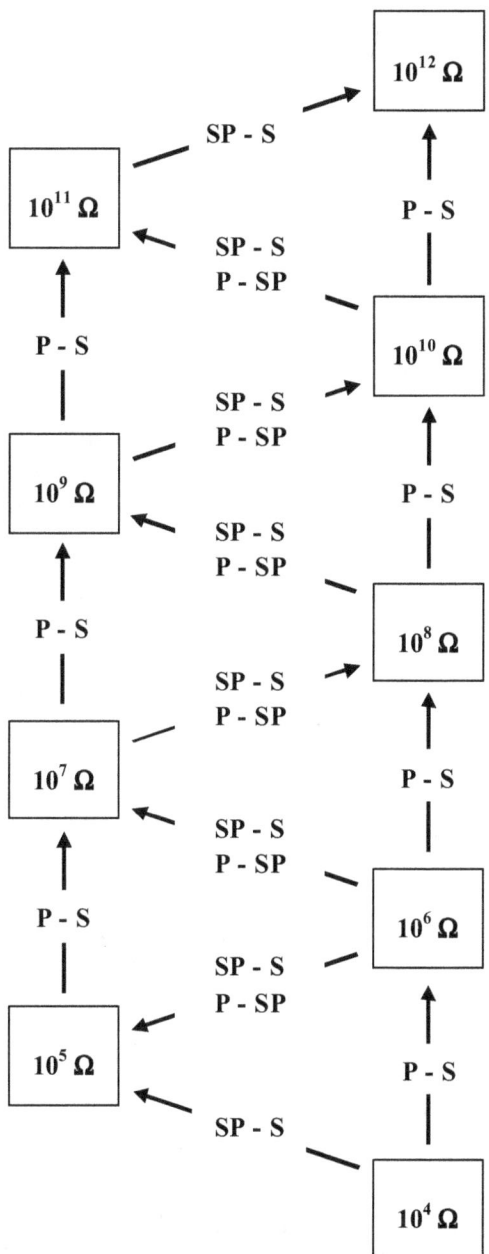

Fig. 8.7. Resistance scaling using Hamon devices. Arrows indicate resistance transfers between decade levels made by reconfiguring a Hamon device.

8.6 Active-Arm Bridge Ratios

Active-arm bridge ratios greater than unity are compared to transfer device ratios at resistance levels between 1 MΩ and 1 TΩ. This data can be used to extend the range of high resistance measurements to 10 TΩ and 100 TΩ. Presently all measurements above 1 TΩ are performed as special tests.

9. MEASUREMENT UNCERTAINTY

The expanded or overall uncertainty assigned to the measured value of resistance of standard resistors, as given in a NIST Report of Calibration, is equal to

$$U = 2\,u_c = 2\sqrt{\sum u_i^2} \qquad (9.1)$$

where u_i is the estimated relative standard uncertainty associated with the ith source of error and expressed as an equivalent standard deviation. One component is an estimated Type A standard uncertainty based on the pooled standard deviation from a large population of individual measurements. The coverage factor 2 used at NIST is consistent with international practice [33]. The method of combining uncertainties is based on the approach recommended by the International Bureau of Weights and Measures (BIPM) [34]. The reported expanded uncertainty contains no allowances for the long-term drift of the resistor under test, for the possible effects of transporting the standard resistor between laboratories, nor for measurement uncertainties in the user's laboratory. The appendix contains samples of some NIST Reports of Calibration.

The calibration of a resistor involves the comparison of the test resistor with NIST working standards based on the Ω_{NIST} using one of several measurement systems previously described. The critical elements contributing to the measurement uncertainty are the standard resistors (working, transfer, check, and test), and the measurement systems.

9.1 Model - Standard Resistors

Experimental data indicate that the change in resistance of a standard resistor is a function of environmental parameters, time, and load level. To some degree, the environmental parameters can be controlled and/or measured in order to correct for their influence. Assuming the resistor is not subjected to mechanical disturbances, the environmental parameters of main concern that contribute to the measurement uncertainty are temperature, pressure, and humidity.

The gradual change of resistance with time cannot be controlled and it is difficult to predict. This characteristic is inherent in a resistor as a result of its design and construction. The four major sources of this resistance change are metallurgical processes, oxidation, bending stresses and strains, and tensile stresses and strains. The combination of these effects results in a unique stability curve for each resistor. Although such curves may exhibit a shape of an exponential decay function, evidence indicates that over the short term the stability curves of resistors can be modeled as linear functions. This evidence includes the results of the absolute determinations of resistance in SI units and, more recently, the monitoring of resistance by the quantum Hall effect by NIST and other national laboratories. In addition, the run-sequence plots of NIST working and check standard resistors over a time span of several years are fitted best by straight lines having slopes of different signs as well as different magnitudes. It is inferred that for standard resistors not mechanically disturbed, it is likely they will drift linearly, and with sufficient data their value can be predicted to a high degree of certainty using a linear least-squares analysis.

The change in resistance of a standard resistor as a function of measurement current or loading, described by the load coefficient of resistance (LCR) or voltage coefficient of resistance (VCR), results from: 1) the i^2R loading or self-heating of a resistor, and/or 2) the voltage dependence of a resistor due to dielectric effects, surface effects, or other processes. The former is primarily a function of the TCR and the surface area of the resistance material, while the latter is significant for film-type resistors, many of which exhibit large VCR's. The absolute values of VCR's for film-type resistors can be as large as 100 $(\mu\Omega/\Omega)/V$ or more.

The change in resistance as a function of temperature is usually the most significant environmental influence factor. The resistance vs. temperature curve for resistors over the temperature interval 20 °C to 30 °C can be represented by

$$R(P) \approx R'[1 + \alpha(T - T_r) + \beta(T - T_r)^2] \quad ,$$

where $R(T)$ is the resistance at temperature T, R' is the resistance at a reference temperature T_r, α is the slope of the curve at T_r, and β is the second-order coefficient that best represents the resistance behavior over the temperature interval as determined from a least-squares fitting procedure. NIST uses a reference temperature T_r of 25 °C for oil-type resistors and 23 °C for air-type resistors. The coefficients α and β do not change appreciably with time.

Some standard resistors (Thomas 1 Ω and air-type special 10 kΩ, for example) exhibit changes in resistance with changes in pressure. From a study of a number of these resistors, it appears that the effect of pressure on resistance is linear over the range of barometric pressure near 100 kPa. A pressure coefficient of resistance (PCR) can be determined can be expressed as

$$R(P) \approx R''[1 + \gamma(P - P_r)],$$

where $R(P)$ is the resistance at pressure P, R'' is the resistance at a reference pressure P_r, and γ is the PCR. The reference pressure is taken to be 101.325 kPa and the PCRs of most Thomas-type resistors lie between $+0.002 \times 10^{-6}$/kPa and $+0.025 \times 10^{-6}$/kPa. For a user located at an altitude of one mile above sea level, the resistance of this type of resistor can be as much as 0.5 ppm less than the value at the altitude of the NIST Gaithersburg laboratory.

Besides the barometric or surface pressure one has to consider the additional pressure effect on the resistor due to its depth of immersion in the oil bath. As an example, for mineral oil with a specific gravity of 0.86 the value of P increases 8.48 Pa for each millimeter descent below the surface of the oil. At NIST the reference position of the resistor is taken to be at the level of the electrical contact surfaces of the current terminals. The value of pressure given in a NIST report of calibration is the sum of the ambient barometric pressure at the time of calibration plus the pressure resulting from the depth of the resistor below the surface of the oil.

9.2 Model - Measurement Systems

The model for the measurement system is broken down into the following main components: the ratio instrument; dummy resistor; detector circuit; procedures and conditions of test; and auxiliary equipment. The measurement technique is usually that of a 1-to-1 comparison, in which a test resistor or check standard is substituted by a working standard of the same nominal value (substitution technique). The ratio instrument is then relied upon to determine only the difference between resistors of the same nominal value, where this difference is relatively small. Stability and linearity of the ratio instrument, leakage resistance at the test resistor terminals, stability of the dummy resistor (if used), detector resolution, and changing thermal EMFs in the detector circuit can significantly affect the measurement. Of secondary importance are errors resulting from other leakage resistance paths, lead resistances, contact resistances and test conditions, which all are greatly reduced when using the substitution technique or a current comparator technique. All other sources of error associated with procedures and auxiliary equipment are believed to be negligible in the systems currently at use at NIST.

The 1 Ω DCC potentiometer system and the 10 kΩ Warshawsky bridge system will be examined in detail. High quality 1 Ω and 10 kΩ resistors have been produced by several manufacturers, and as a consequence, most NIST resistor calibrations with the smallest NIST uncertainties are done at these two resistance levels. First, to establish whether or not the measurement systems are "in statistical control," the data histories of the measurement processes are examined. Figure 9.1 is the control chart for the NIST 1 Ω check standard C84 from 1990 to 2003. Each open-diamond symbol represents a single measurement result taken over a period of a few hours.

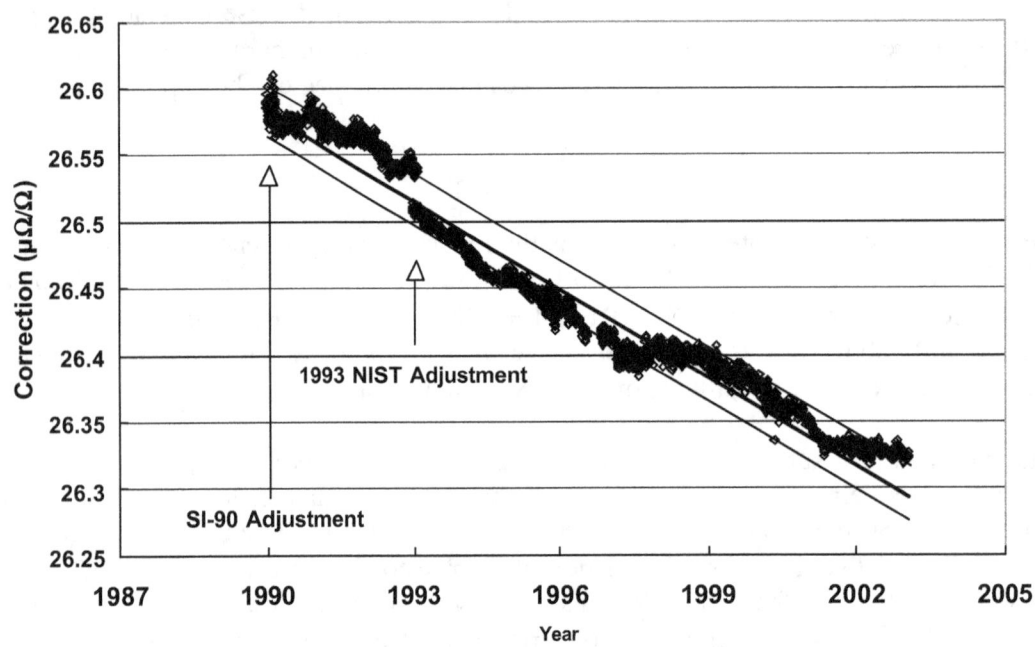

Figure 9.1. Control chart for 1 Ω check standard C84.

The two lighter parallel lines in Fig. 9.1 represent the NIST allowance of ± 0.02 ppm based on the predicted average value of the five NIST 1 Ω working standards, relative to the SI ohm. This allowance is one component of the expanded uncertainty in a NIST Report of Calibration. The check standard C84 is of similar quality to the five NIST 1Ω working standards, and Fig. 9.1 illustrates that this Thomas-type standard (or the working standards) may have somewhat non-linear drift over periods of weeks to years.

In 1991-92, NIST determined that predictions based on the drift factors of the 1 Ω bank of working standards were diverging from values found by scaling to the QHR standard using the CCC method. On January 1, 1990, the numerical values of all NIST reference resistors had been decreased by 1.69 ppm to agree with SI-90. This major adjustment of the representation of the ohm is discussed in earlier NIST publications [2,13]. Three years after the 1990 adjustment, new drift factors predicting the individual values of the resistors making up the 1 Ω bank were required. These new drift factors caused the average value of the five working standards to decrease on January 1, 1993 by 0.03 ppm, and changed their average drift rate by 0.01 (μΩ/Ω)/year. Since 1993, the drift factor predictions have averaged within ± 0.01 ppm of the QHR scaling values for the average of the five Thomas 1 Ω working standards. A trend chart showing the difference between the scaling value (based on the QHR) and the working value (based on the drift factors in use at the time) is shown in Fig. 9.2. Data based on QHR Hamon scaling comparisons are shown as filled diamonds, and CCC scaling measurements are shown as filled squares.

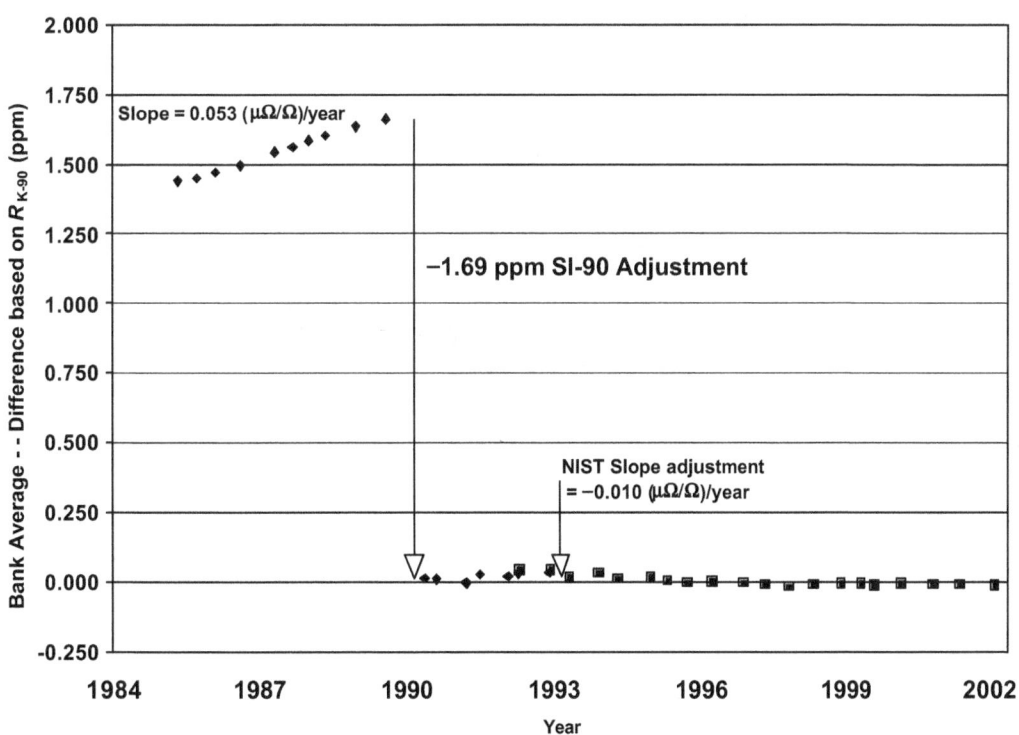

Fig. 9.2. Trend chart for 1 Ω Thomas bank of working standards.

As of late 2002, the average of 26 QHR-based values for the 1 Ω bank of working standards since 1990 (shown in Fig. 9.2) differs from the average value predicted from the drift factors by +0.006 ppm. Since the 1993 correction, the average value for 18 comparisons differs by +0.002 ppm, and for the most recent three-year period, the average results differ by −0.007 ppm. To improve the uncertainty of calibration values, NIST now applies a correction to 1 Ω Thomas-type calibration results based on the most recent 1-2 years' QHR scaling results. This correction works to reduce the small Type B error due to the long-term non-linear drift of the working standards. Completely new drift factors will be applied at such time as the predictions based on the 1 Ω drift factors are in error by approximately ±0.02 ppm.

The NIST 100 Ω resistance bank is the transfer reference used for CCC scaling to working standards at the 100 Ω level and the working standards for the 10 kΩ Warshawsky bridge, as described in section 8.2. Using the CCC, upon request NIST can provide special low-uncertainty calibrations for customer's 100 Ω or 10 kΩ standard resistors if those standards are of a quality equal to the 10 kΩ "special" resistors measured on the Warshawsky bridge.

In 1996-1997, comparisons were made between the manual double-Kelvin type bridge [13] and the automated Warshawsky bridge. The residual standard deviations of the linear least-squares analysis of the data for the manual and automatic systems were 0.008 μΩ/Ω and 0.004 μΩ/Ω, respectively [14]. The 10 kΩ measurement system has one air-type check standard, C1410, and one oil-type check standard, C1419. Figure 9.3 shows control chart data on C1410. Periodically the drift factors of the two working standards are recalculated to improve the agreement with the most recent QHR-based scaling results, and the effect of one such change (Feb. 2002) for the working standards is seen in Fig. 9.3.

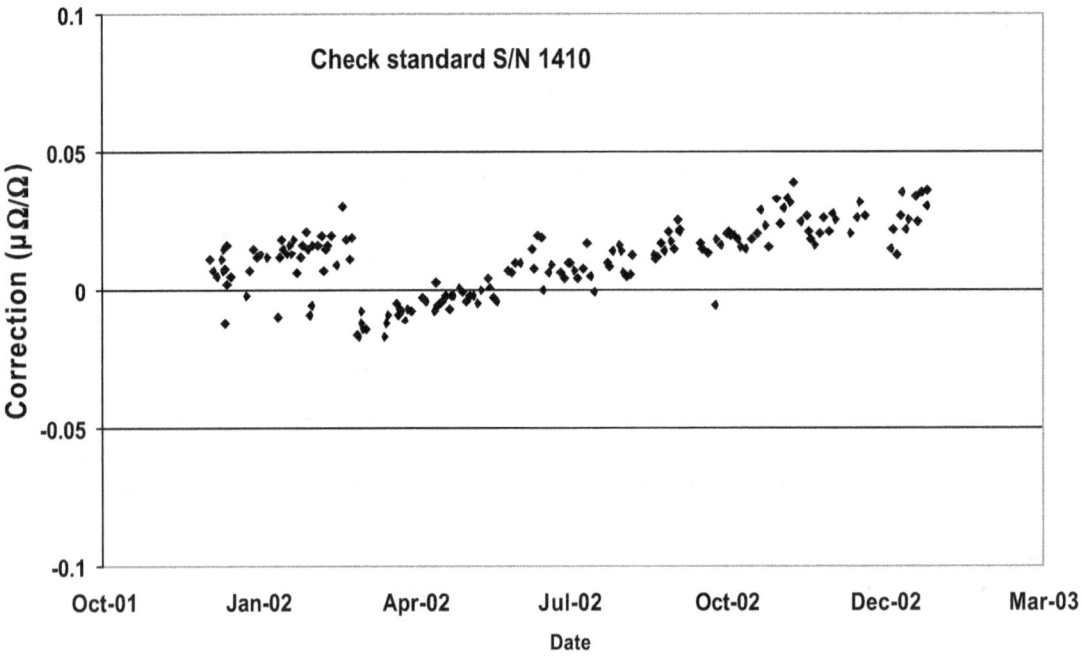

Fig. 9.3. Control chart for 10 kΩ check standard C1410.

9.3 Type A Standard Uncertainty

The Type A standard uncertainty, u_i, assigned to the value of a standard resistor is based on an estimate of the standard deviation of the measurement process for that particular resistance level. This u_i is calculated from a sampling of a large number of measurement runs. First, the individual estimates of the standard deviations s_i of one or more high-quality test resistors are calculated from repetitive measurements taken during a single test run. If more than one test resistor is measured during a run, a pooled value of the estimate of the standard deviation is obtained for that particular run. Finally, the pooled values for at least 15 measurement runs, over a several month interval, are themselves pooled to obtain a realistic estimate of this u_i for the measurement process.

The Type A measurement uncertainty is a small contribution to the combined uncertainty for nearly all present-day NIST resistance calibrations. In the calibration of a customer standard, special allowance is made only if the statistical uncertainty component of the customer's result significantly exceeds the Type A pooled value result. Generally, excessive variation in the measurement data is in fact attributable to Type B influences, as defined in section 9.4.

Each customer standard is measured over a period of approximately two weeks, the results are plotted, and the drift and standard deviation are compared to typical results. If the drift in the measured resistance is greater than expected, for example if significantly more drift is observed during the initial week's results, the measurement period may be extended and only the last two week's results would be used in the calibration average. Careful packing and transportation of standard resistors is necessary to reduce the possibility of excessive mechanical shock or large temperature changes, which may cause permanent changes in the resistance value or drift rate.

9.4 Type B Standard Uncertainty

The Type B standard uncertainty assigned to the value of a standard resistor is determined by combining in quadrature (root-sum-square or RSS) the estimated standard uncertainties from each known individual component of the measurement process for that particular resistance level. Each Type B standard uncertainty, u_i, is either equal to $1/(2\sqrt{3})$ times an estimated range for a uniform distribution, or a similar estimate. For analysis, the sources of error for the Type B standard uncertainties are separated into two main categories: 1) those associated with the standard resistors, and 2) those associated with the measurement systems. Tables 9.1 and 9.2 list the estimates of the Type B standard uncertainties for the various resistance levels. No allowance is made in the Type B uncertainty for the effects of transportation, drift, or environmental influences before or after the resistor arrives at the NIST laboratory for testing.

9.4.1 Standard Resistors

This category includes the main sources of error associated with the Type B standard uncertainties assigned to the values of the working standards. These sources of error are subdivided into three main areas: a) maintaining the working Ω_{NIST} in terms of the quantized Hall resistance, b) scaling from the Ω_{NIST} to working standards at other resistance levels using

CCC bridges or transfer standards, and c) the environmental and drift factors of the working standards themselves.

a) NIST ohm – The 100 Ω reference group used with the QHR and the 1 Ω working group of five Thomas-type resistors are used to preserve the Ω_{NIST} between QHR comparisons. Since 1992 the 1 Ω working group's link with the QHR has been maintained using the 129.06-to-1 and 100-to-1 CCC ratios. Recent references [30, 35] list in detail the Type A and Type B standard uncertainties for this process. From these results, a conservative estimate of the relative uncertainty of resistance scaling based on the QHR standard is 0.01 ppm or one part in 10^8.

b) Primary scaling (CCC bridges and transfer standards) – Scaling to the 1 Ω and 10 kΩ levels has been discussed in section 9.2. Recently, scaling directly from the QHR to the 1 MΩ level using a 77.48-to-1 CCC ratio has been used to derive improved values of working standards at 1 MΩ and higher resistance levels. Accurate transfer standard ratios of 10-to-1 and 100-to-1 are also necessary for extending the Ω_{NIST} to working standards at higher resistance levels. Under this listing in Tables 9.1 and 9.2 are the combined standard uncertainties of all the scaling methods used to assign a value, based on Ω_{NIST}, to a particular working standard.

Included in the Type B uncertainties listed in Tables 9.1 and 9.2 are estimates of the ratio errors of the various Hamon-type transfer standards for resistance levels at 100 kΩ and at 10 MΩ and above. An estimate of the ratio error of a particular transfer standard can be calculated from measurements of: 1) the differences among the main resistors, 2) the finite resistances of the 4-terminal junctions, and 3) the resistances of the paralleling fixtures including fan resistors. If loading errors are small, the 10-to-1 ratio can be verified by comparing the results obtained from the series-parallel measurements as described in section 8.2 against the mean value of the main resistors obtained by measuring each individual resistor. Leakage resistances within transfer standards of high resistance can cause significant ratio errors. Surface leakage at the connectors is eliminated by guard resistor networks with isolated guards at the terminations of these transfer standards.

Loading effects are also a part of primary scaling uncertainty for DCC and CCC ratios included in Tables 9.1 and 9.2. These effects are small for the 1 Ω and 100 Ω levels, because scaling is performed at a power dissipation level of 10 mW for these standards, at the same loading as calibration measurements. The LCR of the 10 kΩ resistors used in CCC scaling is possibly significant and selected 10 kΩ resistors having small LCR for loads between 0.1 mW and 10 mW are used in scaling.

Loading effects have been carefully evaluated for the 100 Ω resistor bank. The low average LCR of these standards allows them to be used at different power levels up to 10 mW with negligible change in resistance. Thomas 1 Ω scaling measurements and customer calibration measurements all take place with 10 mW dissipated in the 1 Ω resistors. At the 1 Ω level, loading effects due to the duty cycle (or ramped current reversal in the CCC measurements) are negligible because the 1 Ω transfer standards used in scaling have very low TCR and LCR coefficients. NIST measurements for resistor scaling and calibrations from 1 Ω to 1 MΩ usually have small uncertainties due to loading since the measurements are done at a power level of 10 mW or less, and resistors used as working standards have low loading coefficients (LCR ≤ 1 (µΩ/Ω)/W).

c) Working Standards – A set of working standards exists at each decade resistance level. The Type B standard uncertainty assigned to a particular set of working standards is based, primarily, on the knowledge of the drift factors and also on environmental parameters that affect the set of standards during a measurement run.

9.4.2 Measurement Systems

Listed in Tables 9.1 and 9.2 are the estimated Type B standard uncertainties for the various sources of systematic error associated with the measurement systems for each resistance level. The sources of error are subdivided into five classifications as follows:

a) Ratio – Linearity of ratios is ensured by the design of the measurement systems where substitution techniques are used. For example, if the test resistor and working standard values differ by 100 ppm, the difference need only to be determined to 1% to achieve a 1 ppm relative measurement of the test resistor based on the working standard.

More than one working standard is used in all calibration systems at 1 MΩ and below, and check standards are used in all systems except with some high-current shunts, where check standards often are not available. Above 1 MΩ, scaling between levels is based on Hamon type transfer standards, and these are used as the primary working standards at most resistance levels.

DCC bridge ratios greater than unity are compared in one or more steps to CCC ratios. The manual DCC bridge has provisions for self-calibrating its 10-to-1, 100-to-1, and 1000-to-1 ratios. The 1000-to-1 range-extender ratio of the DCC automatic system and the manual DCC 2000 A range extender ratio have been compared and agree to better than 5 parts in 10^6 for high-current shunt measurements above 300 A. At lower currents, the ratios of manual DCC bridges and the DCC automatic systems at NIST agree to better than one part in 10^7 for ranges up to 100-to-1.

b) Stability – These include the errors caused by the instability of the ratio arms of a resistance bridge, instability of the thermal EMF in the circuit, and by the short-term instability of a dummy resistor in a bridge or DCC measurement system. DCC bridges inherently have very stable ratios and are largely immune to these errors. In resistance-ratio bridges, one cause of uncertainty is the change in the lead and contact resistances of a network during a test run. Usually, resistance-ratio bridge short-term instability results from changes in these network properties or environmental conditions during a measurement run. Reduction of these errors is achieved when reference standards are measured both at the beginning and end of a measurement run in NIST substitution type measurements (DCC potentiometer, Warshawsky bridge) or at multiple points throughout the measurement run (ring method, active-arm bridge).

c) Detector – The error in the detector circuit is associated with the stability of the zero setting of the detector and the resolution of the detector system. For the automated DCC potentiometer and DCC bridge systems, it also includes the errors caused by the non-linearity of the detector and feedback circuits.

d) Ambient – This category contains the uncertainty in resistance due to: 1) the estimated error of the internal temperature of the customer's resistor and the working standards from the indicated temperature, 2) the estimated error of those temperatures as measured from the true temperature as derived from the ITS-90, 3) the estimated error of the pressure measurement for a Thomas-type 1 Ω or air-type 10 kΩ measurement, and 4) the estimated error due to uncertainty in humidity measurement for the high resistance measurements. The critical temperature measurements are made using a calibrated PRT. The secondary temperature measurements are made with digital thermistor thermometers that are calibrated against a PRT. The TCR of a customer's resistor is often not known, but it can be estimated from that of similar standards.

e) Leakage – This uncertainty results from the possible deterioration of the insulating properties of the mineral oil and/or cables and connectors. The insulation of the mineral oil in the baths is checked periodically and the oil is changed if the oil resistance (between opposite terminals of the resistance standards) falls below 5×10^{13} Ω. Lead connections between the resistors and the other parts of the bridge are made using shielded, PTFE-insulated cable with silver-plated copper conductors. The shields of these cables and BPO connectors are driven at a guard potential if the system uses an auxiliary guard network. However, because of the many exposed conductors present in the Ring method system, at higher resistance leakage can be a significant source of uncertainty. We have observed a positional effect for 1 MΩ resistance standards, where the result of calibration depends on the relative position of the working standards and other standards, and have developed an alternate measurement system to reduce this source of uncertainty.

NIST has begun using a guarded Warshawsky bridge for oil-type and air-type 1 MΩ resistance standards, because the leakage resistance paths are greatly reduced relative to the Ring Method system. CCC-based validation of the Warshawsky bridge will be complete by the time of this publication. By special arrangement, air-type customer standards can also be measured at the time when NIST calibrates its working 1 MΩ standards based on the QHR. In this case, we can assign a substantially reduced total uncertainty at 1 MΩ.

All calibrations at 10 MΩ and above are made in air, at 35% ± 5% relative humidity, to reduce leakage paths. Hermetically sealed resistance units like those constructed at NIST have conducting surfaces that intercept and/or reduce leakage currents and ambient effects. A hermetic shield reduces the effect of ambient humidity on insulator surfaces and dielectric coatings. Since these effects are of substantial benefit in reducing measurement uncertainty, NIST offers a special class of calibration uncertainties in the high-resistance range for suitable hermetically sealed standards.

Table 9.1 Type B uncertainty influence factors (in ppm) for lower resistance ranges.

Resistance	Standards				System				
(Ω)	Ω(NIST)	Primary	Working	Ratio	Stability	Detector	Ambient	Leakage	Connection
1(T)	0.010	0.005	0.010	0.002	0.002	0.002	0.004	0.001	0.002
10^4(S)	0.010	0.020	0.020	0.005	0.005	0.002	0.005	0.005	0.005
10^{-4}	0.01	0.02	0.05	0.20	0.50	1.0	0.10	0	1.0
10^{-3}	0.01	0.02	0.05	0.10	0.20	0.30	0.10	0	0.30
10^{-2}	0.01	0.02	0.05	0.10	0.10	0.10	0.10	0	0.20
10^{-1}	0.01	0.02	0.05	0.10	0.05	0.05	0.10	0	0.15
1	0.01	0.01	0.05	0.02	0.05	0.05	0.06	0	0.10
10	0.01	0.02	0.05	0.02	0.05	0.05	0.06	0	0.05
10^2	0.01	0.01	0.05	0.02	0.02	0.02	0.06	0	0.01
10^3	0.01	0.02	0.05	0.02	0.02	0.02	0.06	0	0
10^4	0.01	0.02	0.05	0.02	0.02	0.02	0.06	0.01	0
10^5	0.01	0.05	0.20	0.10	0.20	0.05	0.06	0.05	0
10^6	0.01	0.05	0.10	0.10	0.20	0.20	0.06	0.20	0

Table 9.2 Type B uncertainty influence factors (in ppm) for high resistance ranges.

Resistance		Standards			System					
(Ω)	Ω(NIST)	Primary	Working	Ratio	Stability	Detector	Ambient	Leakage	Repeatability
10^7	0.01	0.10	0.20	1.0	0.10	0	2.0	0.02	2.0
10^7(S)	0.01	0.10	0.20	1.0	0.10	0	0.50	0.02	0.10
10^8	0.01	0.10	0.50	1.0	0.10	0	2.0	0.05	5.0
10^8(S)	0.01	0.10	0.50	1.0	0.10	0	1.0	0.05	0.30
10^9	0.01	0.50	2.0	1.0	0.50	0.1	5.0	0.20	10
10^9(S)	0.01	0.50	2.0	1.0	0.50	0.1	2.0	0.10	1.0
10^{10}	0.01	0.50	5.0	1.0	0.50	0.5	5.0	2.0	20
10^{10}(S)	0.01	0.50	5.0	1.0	0.50	0.5	2.0	1.0	3.0
10^{11}	0.01	2.0	10	1.0	2.0	2.0	10	5.0	50
10^{11}(S)	0.01	2.0	10	1.0	2.0	2.0	5.0	2.0	10
10^{12}	0.01	2.0	20	1.0	2.0	20	20	50	100
10^{12}(S)	0.01	2.0	20	1.0	2.0	20	10	20	20

9.4.3 Measurement Repeatability

Some special influence factors that limit the repeatability of measurements may contribute to the quoted uncertainty. When NIST measurements show that such additional influence factors are present, we follow guidelines as described below to calculate an additional Type B uncertainty term. These Type B terms, if necessary, may be determined for each resistor at the time of test.

a) Connection effect – This includes the estimated errors associated with: 1) the location of the potential points for a Rosa-type standard resistor, 2) the effects of making connections with automatic scanners, especially for high resistance measurements, because of different dielectric properties in the scanner circuits and cables, and 3) the possible variation in current distribution when the current connections are relatively close to the potential connections, especially in low-value and high-current shunt resistors. The resistivity of the terminal arms of a commercial Rosa-type standard resistor is about 3.5 $\mu\Omega$/cm. The maximum error of positioning a potential lead to a screw-type terminal is about 0.5 cm for each arm. Therefore, for Rosa type resistors of 10 Ω or below, permanent stand-off connectors are affixed at the potential connection points to allow more repeatable potential connections. Even so, additional uncertainty to that given in Table 9.2 results for Rosa-type standards of nominal value below 1 Ω.

b) High-voltage settling effects – High value film-type resistors normally have a voltage coefficient of resistance related to dielectric changes or other changes when a voltage is applied. A settling time (typically 120 s or 300 s, depending on the nominal resistance and the quality of the standard) after each current reversal is used in NIST measurements to reduce the effect of these changes on the data. Excessive settling effects can be characterized by increasing the settling time and noting if there is a significant change in the measured resistance value.

c) Hysteretic influences – Changes due to ambient or applied conditions that affect resistance (temperature, humidity, pressure, current or voltage) may be reversible but long-lasting. High-value resistors (both wire and film types) are likely to have hysteretic influences due to humidity or high-voltage surface film effects because of their comparatively large surface area per unit volume of resistance material. Hermetically sealed high-value resistance elements are better protected from these influences. The construction, type of resistance element, and voltage coefficient of high-value resistance standards help to indicate the possible magnitude of these effects. Typically, hysteretic behavior is characterized by a larger than normal statistical distribution of results in some fraction or segment of the measurements.

d) Medium-term repeatability – NIST routinely takes into account environmental conditions such as large changes in barometric pressure for resistors with large PCR, and changes in temperature for air-type standards. However, some standard resistors may exhibit unexplained significant changes in value during the measurement period at NIST. This may occur temporarily as the result of a mechanical disturbance or thermal shock during transport. If the instability is still noteworthy after one to two weeks, the reported uncertainty may be increased. Occasionally, significant changes in resistance value or a resistance value that is far from nominal may indicate that the mechanical construction of the standard has been compromised. In these cases, the owner of the resistor will be contacted before a report is issued.

9.5 Combined Uncertainty

Table 9.3 shows the Type A standard uncertainty, the total RSS contribution of Type B standard uncertainty, and the resulting relative combined (RSS total) uncertainty for each resistance level from 10^{-4} Ω to 10^{12} Ω.

9.6 Reported Uncertainty

Table 9.4 shows the expanded uncertainties for standard resistors, as given in a NIST Report of Calibration. These uncertainties are based on calculations using eq. (9.1) with a coverage factor of 2 and the estimated Type A and Type B standard uncertainties listed in tables 9.1 through 9.3. In most cases the reported uncertainties have been conservatively rounded upward. Table 9.4 lists the maximum power dissipation of a test resistor during calibration, except for resistances of 10^7 Ω or above where the customer specifies the test voltage. Also listed are the test numbers for each of the resistance levels.

Some resistance standards are designed in such a way that under normal use, influence factors specific to the resistor under test such as drift, leakage, the response to ambient conditions, and certain other factors listed in this section have been greatly reduced. This includes Thomas-type or equivalent 1 Ω standards, two types of 10 kΩ standards, and some high-value resistance standards of wire-wound or hermetically-sealed construction. These types of standards are measured at lower uncertainties and a capital letter (T) or (S) is added to the value in the resistance columns in Tables 9.2 and 9.3. In the high resistance range, test numbers are the same for normal (open, film-type) and special (wire-wound or hermetically-sealed) types of resistors. For all high-value resistors, the measurement voltage also must be at least 10 V for the standard to be given the lower uncertainty.

Table 9.3 Standard uncertainties (in ppm) for low-power standard resistors.

Resistance (Ω)	Type A (u_0)	Type B $\sqrt{\sum_{i \neq 0} u_i^2}$	RSS Total $\sqrt{\sum u_i^2}$
1(T)	0.005	0.016	0.017
10^4(S)	0.005	0.032	0.032
10^{-4}	0.70	1.52	1.67
10^{-3}	0.30	0.49	0.58
10^{-2}	0.15	0.29	0.33
10^{-1}	0.05	0.23	0.23
1	0.02	0.15	0.15
10	0.02	0.12	0.12
10^2	0.02	0.09	0.09
10^3	0.02	0.09	0.09
10^4	0.02	0.09	0.09
10^5	0.04	0.32	0.32
10^6	0.10	0.38	0.40
10^7	0.5	3.1	3.1
10^7(S)	0.5	1.2	1.3
10^8	0.5	5.5	5.5
10^8(S)	0.5	1.5	1.6
10^9	1	11.4	11.5
10^9(S)	1	3.2	3.4
10^{10}	5	21	22
10^{10}(S)	5	6.7	8.1
10^{11}	10	52	53
10^{11}(S)	10	16	19
10^{12}	25	117	120
10^{12}(S)	25	41	48

Table 9.4 Low-power standard resistor expanded uncertainties with coverage factor $k = 2$ based on Table 9.3.

SP-250 [Ref. 4] Test No.	Resistance (Ω)	Power (mW)	Uncertainty U (ppm)
51130C	1(T)	10	0.04
51131C	10^4(S)	10	0.08
51132C	10^{-4}	10	4
51133C	10^{-3}	10	1.2
51134C	10^{-2}	10	0.8
51135C	10^{-1}	10	0.5
51136C	1	10	0.3
51137C	10	10	0.3
51138C	10^2	10	0.2
51139C	10^3	10	0.2
51140C	10^4	10	0.2
51141C	10^5	10	0.8
51142C	10^6	1.1	0.8
51143C	10^7	*	8
	10^7(S)	*	3
51145C	10^8	*	12
	10^8(S)	*	5
51147C	10^9	*	25
	10^9(S)	*	10
51149C	10^{10}	*	50
	10^{10}(S)	*	20
51151C	10^{11}	*	120
	10^{11}(S)	*	50
51153C	10^{12}	*	250
	10^{12}(S)	*	100

9.7 Current Shunt Uncertainties

Shunt resistors used for measurements of high current differ from low-power standard resistors. Current shunt resistors are constructed by many different designs. The process of determining Type B uncertainty is difficult since power-related influences are dependent on the design. The type and location of potential and current connections contribute to uncertainty in these calibrations. Current-lead connections should be made with appropriate heavy connectors and utilize all bolt-type connection points that are provided. NIST staff evaluate any loading and connection effects and estimates the Type B uncertainty at the time of test.

For high current standard resistors under load, air flow over the resistance element causes the resistance to reach equilibrium at a lower temperature, compared to a measurement in still air. An increased uncertainty results for high load conditions because the equilibrium resistance depends on the air velocity and other factors specific to the NIST laboratory at the time of measurement. If a copper-constantan thermocouple is permanently attached to the resistance element, NIST can measure the temperature of that sensor during the calibration. When thermocouple sensor temperature data is recorded an equilibrium temperature for the sensor will be given on the NIST calibration report.

After 120 min under load, most high-current shunt resistors are close to thermal equilibrium. NIST staff examine the change in resistance after current is applied to a shunt resistor, beginning at either 10 min (for currents up to 100 A) or 30 min (for currents above 100 A) after the current is applied, and ending when the resistance value has reached approximate equilibrium. The difference in resistance value from the beginning to the end of this period is used to assign the loading repeatability uncertainty. When this influence factor is appreciable, NIST provides the owner of the shunt resistor with one or more typical graphs showing the change of resistance with time (and temperature if a copper-constantan/type-T thermocouple is provided) over the measurement period. If the influence is correlated with the temperature indicated by the thermocouple sensor (as is usually the case), this Type B uncertainty term generally is reduced.

NIST calibrates both individual units and combination or "multi-range" shunt resistor sets. Table 9.5 gives the estimated Type B uncertainty budgets for typical multi-range current shunt resistors, consisting of nine shunt resistors (3.333×10^{-4} Ω to 10^4 Ω) packaged in one enclosure. The design of these shunts reduces uncertainty due to connections and the enclosure helps to eliminate air currents that affect repeatability. Each of the nine resistors are measured with the current level set to produce a voltage drop of approximately 100 mV across the potential terminals.

Table 9.6 gives Type A, Type B, and combined expanded uncertainties for some NIST current shunt calibrations. Type A and Type B uncertainties given in Table 9.6 are typical, but actual uncertainties are determined at the time of test based on the construction, power rating, use of an attached thermocouple, and the behavior of the current shunt resistor.

Table 9.5 Type B uncertainty (in ppm) for typical multi-range current shunt resistor sets.

Resistance (Ω)	Power (W)	Standards (ppm)	Ratio	Stability	System Detector	Ambient	Loading
3.333×10^{-4}	30	0.20	0.80	0.40	1.50	3.0	10
10^{-3}	10	0.10	0.40	0.40	0.50	3.0	8.0
10^{-2}	1	0.10	0.20	0.20	0.50	1.5	2.0
10^{-1}	10^{-1}	0.10	0.10	0.10	0.10	1.5	0.50
1	10^{-2}	0.10	0.02	0.05	0.05	1.5	0.10
10	10^{-3}	0.10	0.05	0.05	0.05	1.5	0
10^{2}	10^{-4}	0.10	0.02	0.05	0.05	1.5	0
10^{3}	10^{-5}	0.10	0.05	0.10	0.05	1.5	0
10^{4}	10^{-6}	0.10	0.05	0.50	0.50	1.5	0

Table 9.6 Typical expanded uncertainties with coverage factor $k = 2$ for multi-range and high-current shunts.
  * Applies to typical multi-range current shunt resistors

Resistance (Ω)	Power (W)	Type A (ppm)	Type B (ppm)	Uncertainty (ppm)
1×10^{-5}	10 - 20	5.0	20	50
	>50	5.0	50	100
1×10^{-4}	5 - 20	5.0	10	25
	>50	5.0	20	50
3.333×10^{-4}	30	2.0	10.5	25*
1×10^{-3}	10	2.0	8.6	20*
	>20	2.0	20	50
1×10^{-2}	1	2.0	2.6	7*
	>10	1.0	10	25
1×10^{-1}	10^{-1}	0.50	1.6	4*
	>10	1.0	10	25
1	10^{-2}	0.05	1.5	3*
10	10^{-3}	0.05	1.5	3*
10^{2}	10^{-4}	0.05	1.5	3*
10^{3}	10^{-5}	0.05	1.5	3*
10^{4}	10^{-6}	0.50	1.7	4*

10. QUALITY CONTROL

It is important that procedures are in place to maintain the measurement process in a state of control if the uncertainty that NIST assigns to the value of a standard resistor is to be meaningful. Three factors are examined for each calibration performed in order to ensure the quality of the measurement value that is assigned to a standard resistor:

1. Standard deviations of the measurement runs.
2. Control charts of check standards.
3. Data history of the standard resistor under test.

The computer record of the data analysis/results contains the individual standard deviations, s_i, calculated from the m repeated measurements of sample size n for the resistor over a period of approximately two weeks. The standard deviation for these results is compared to the Type A standard uncertainty as listed in Table 9.3 using the following analysis, which is found in many textbooks on statistical quality control [36].

The average of the m standard deviations is

$$\bar{s} = \frac{1}{m}\sum_{i=1}^{m} s_i \quad ,$$

which is used as an estimator of the population standard deviation of a resistor (s).

We make use of the factor c_4 described in [36], where c_4 is a constant that depends on the sample size n. This constant is calculated using

$$c_4 = \sqrt{\frac{2}{n-1}} \frac{\left(\frac{n}{2}-1\right)!}{\left(\frac{n-1}{2}-1\right)!} \quad .$$

The factorial function (Gamma function) is defined by the following equation,

$$\left(\frac{n}{2}\right)! = \left(\frac{n}{2}\right)\left(\frac{n}{2}-1\right)\left(\frac{n}{2}-2\right)\cdots\left(\frac{1}{2}\right)\sqrt{\pi} \quad .$$

If the underlying distribution is normal, then the mean or expected value of the sample standard deviation, s as well as $\bar{s}$, is $c_4 s$. The standard deviation of the sample standard deviation is

$$\sigma_s = \sigma\sqrt{1-c_4^2} \quad .$$

Obviously, σ_s can be estimated by $\overline{s}\sqrt{1-c_4^2}/c_4$. Therefore, for a coverage factor $k = 2$, the parameters of the control chart would be

$$Upper\ and\ Lower\ Control\ Limits = \overline{s} \pm 2\frac{\overline{s}}{c_4}\sqrt{1-c_4^2}.$$

If the Type A standard uncertainty of the measurement process is considered out of control, appropriate steps may be taken to correct the situation. Recently, computer automation has largely removed operator error as a source of measurement error. However, electromagnetic interference from office equipment has become more prevalent. Whatever the cause, any identifiable problem is corrected as soon as it is identified and the measurement test is repeated.

For each of the check standards, the predicted value is calculated from drift factors that are derived from a previous least-squares analysis of a linear fit for that check standard. If this difference or residual exceeds the Type A standard uncertainty by a factor of 2, it indicates that a potential problem may exist and that the other control indicators should be scrutinized. The excessive values of the residuals may be caused by a lack of knowledge of the resistor model for specific resistors. The data histories of all standard resistors measured in the automated NIST bridges are stored in a central database and are accessible locally by NIST resistance calibration staff. The central database can be used to quickly generate the control chart for that check standard to determine if it is within acceptable limits.

Finally, the data history of the standard resistor under test is examined if more than two measurement points exist. Additional tests are run to determine the stability of the resistor, and if necessary the customer is apprised of the situation.

11. ONGOING PROJECTS

The staff at NIST endeavors to improve the maintenance and dissemination of Ω_{NIST}, to provide new and improved calibration services for dc resistors where needed, and to make techniques that allow better resistance measurements available to our customers for use in their laboratories. Direct resistance scaling from the QHR standard to 1 Ω, 100 Ω, 10 kΩ, 1 MΩ, and 100 MΩ can now be conducted using four different CCC bridges. The direct QHR-to-1 MΩ, CCC-based scaling uncertainty is less than 0.1 ppm, and the first measurements with the QHR-to-100 MΩ CCC have a type A uncertainty near 0.5 ppm or better.

Near the time of this publication most of the resistance laboratory functions will be relocated to the new NIST Advanced Measurement Laboratory (AML). In preparation, measurement systems have been rebuilt or newly constructed, to allow uninterrupted calibration of Thomas-type 1 Ω resistors and 10 kΩ special Evanohm-type resistors. The duplicate Warshawsky bridge has been tested at 10 kΩ and 1 MΩ, with early results pointing to equivalence with the original NIST Warsharsky bridge for 10 kΩ measurements, and better than 0.2 ppm overall uncertainty at 1 MΩ.

For the most part, the full range of resistance services will reside in a single large laboratory in the AML. A new automated high-current (400 A) commercial DCC system will be installed for calibrations of current shunts and standard resistors between 10 µΩ and 1000 Ω. This DCC system has suitable capability for resistance calibrations at the 10 kΩ level, based on 10-to-1 ratio build-up from 100 Ω and 1 kΩ working standards. The ring method measurement system thus might be replaced by using the DCC method at the 1 kΩ and 10 kΩ levels and the fully guarded Warshawsky bridge at the 100 kΩ and 1 MΩ levels.

NIST can arrange special tests to determine the effect of influence factors such as barometric pressure and internal power dissipation for state-of-the-art standard resistors. Our staff can provide guidance on ways to improve the systems used to monitor and record environmental data (temperature, humidity and pressure) and to measure and control the temperature of oil-baths and air-baths.

12. ACKNOWLEDGMENTS

The authors wish to express their appreciation to recently retired NIST staff who contributed in many ways to the resistance area. This Technical Note is based on TN 1298 (November 1992), written by Ronald F. Dziuba, and much of the additional material originated with Ron (who, with Donald Sullivan, developed the early overlapped-tube cryogenic current comparator). Theodore P. Moore operated the manual DCC bridges in the 1980s and 1990s, and contributed to this work with his expertise in current shunt and other resistor measurements. Our former group leader Norman Belecki and present group leader Mike Kelley have strongly supported the research and development in this area. In addition we thank Denise Prather, our calibration coordinator, for the support she provides both to our customers and to the staff.

13. REFERENCES

[1]   K. v. Klitzing, G. Dorda, and M. Pepper, "New method for high-accuracy determination of the fine-structure constant based on quantized Hall resistance," Phys. Rev. Lett., vol. 45, pp. 494-497, Aug. 1980.

[2]   B. N. Taylor and T. J. Witt, "New international electrical reference standards based on the Josephson and quantum Hall effects," Metrologia, vol. 26, pp.47-62, 1989.

[3]   M. E. Cage, R. F. Dziuba, C. T. Van Degrift, and D. Yu, "Determination of the time dependence of Ω(NBS) using the quantized Hall resistance," IEEE Trans. Instrum. Meas., IM-38, pp.263-269, Apr. 1989.

[4]   "NIST Calibration Services Users Guide," Natl. Inst. Stand. Technol. Spec. Publ. 250, for sale by the Superintendent of Documents, U. S. Government Printing Office, Washington, DC 20402. Similar information is available in the NIST Technology Services web site, at http://ts.nist.gov/.

[5]   P.A. Boynton, J. E. Sims, and R. F. Dziuba, "NIST Measurement Assurance Program for Resistance," Natl. Inst. Stand. Technol. Technical Note 1424, Nov. 1997.

[6] M. P. MacMartin and N. L. Kusters, "A direct-current-comparator ratio bridge for four-terminal resistance measurements," IEEE Trans. Instrum. Meas., IM-15, pp. 212-220, Dec. 1966.

[7] M. P. MacMartin and N. L. Kusters, "The application of the direct current comparator to a seven-decade potentiometer," IEEE Trans. Instrum. Meas., IM-17, pp. 263-268, Dec. 1968.

[8] I. Robinson, J. Williams, and D. Brown, "Comparison of a Measurements International model 6010B current comparator bridge with a cryogenic current comparator at the National Physical Laboratory," Proceedings of the 1999 NCSL Workshop and Symposium, Charlotte, NC, USA, pp. 195-198, July 15-19, 1999.

[9] K. R. Baker and R. F. Dziuba, "Automated NBS 1 Ω measurement system," IEEE Trans. Instrum. Meas., IM-32, pp. 154-158, Mar. 1983.

[10] R. D. Cutkosky, "An automated resistance thermometer bridge," IEEE Trans. Instrum. Meas., IM-29, pp. 330-333, Dec. 1980.

[11] F. W. Wenner, "Methods, apparatus and procedures for the comparison of precision standard resistors," Natl. Bur. Stand. (U. S.) J. Res., RP1323, vol. 25, pp. 229-294, Aug. 1940.

[12] G. D. Vincent and R. M. Pailthorp, "Experimental verification of the five-terminal ten-kilohm resistor as a device for dissemination of the ohm," IEEE Trans. Instrum. Meas., IM-17, pp. 239-244, Dec. 1968.

[13] R. F. Dziuba, P. A. Boynton, R. E. Elmquist, D. G. Jarrett, T. P. Moore, and J. D. Neal, "NIST measurement service for DC standard resistors," Natl. Inst. Stand. Technol. Technical Note 1298, Nov. 1992.

[14] R. F. Dziuba and L. L. Kile, "An automated guarded bridge system for the comparison of 10 kΩ standard resistors," Proceedings, IEEE Instrumentation and Measurement Technology Conference, Ottawa, Canada, pp. 394-396, May 19-21, 1997.

[15] R. F. Dziuba, D. G. Jarrett, and J. D. Neal, "Wide-range automated system for measuring standard resistors," 1988 Conference on Precision Electromagnetic Measurements, IEEE CPEM'88 Digest, pp. 158-159, June 7-10, 1988.

[16] L. L. Kile, "Programmable guarded coaxial connector panel," Proceedings, 1995 NCSL Workshop and Symposium, Dallas, TX, USA, pp.285-289, July 16-20, 1995.

[17] B. L. Dunfee, "Method for calibrating a standard volt box," Natl. Bur. Stand. (U. S.) J. Res., vol. 67C, pp. 1-13, March 1963.

[18] J. M. Cameron, M. C. Croarkin, and R. C. Raybold, "Designs for the calibration standards of mass," Natl. Bur. Stand. (U. S.) Technical Note 952, June 1977.

[19] R.F. Dziuba, D. G. Jarrett, L. L. Scott, and A. J. Secula, "Fabrication of high-value standard resistors," IEEE Trans. Instrum. Meas., vol. 48, pp. 333-337, April, 1999.

[20] B. V. Hamon, "A 1-100 Ω build-up resistor for the calibration of standard resistors, J. Sci. Instr., vol 31, pp 450-453, Dec. 1954.

[21] D. G. Jarrett, "Evaluation of guarded high-resistance Hamon transfer standards," IEEE Trans. Instrum. Meas., vol. 46, pp. 324-328, Apr. 1999.

[22] L. C. Henderson, "A new technique for the automated measurement of high valued resistors," J. Phys., Electron., Sci. Instrum., vol. 20, pp. 492-495, Sept. 1987.

[23] D. G. Jarrett, "Automated guarded bridge for calibration of multi-megohm standard resistors from 10 MΩ to 1 TΩ," IEEE Trans. Instrum. Meas., vol. 46, pp. 325-328, Apr. 1997.

[24] D. G. Jarrett, J. A. Marshall, T. A. Marshall, and R. F. Dziuba, "Design and evaluation of a low thermal electromotive force guarded scanner for resistance measurements," Rev. Sci. Instrum., vol. 70, pp. 2866-2871, June 1999.

[25] F. Delahaye, "Technical guidelines for reliable measurements of the quantized Hall resistance," Metrologia, vol. 26, pp. 63-68, 1989.

[26] Quantized Hall Resistance Recommended Intrinsic-Derived Standards Practice, RISP-3, August, 1997, Published by the National Conference of Standards Laboratories, 1800 30th St., Suite 305B, Boulder, CO, 80301.

[27] R. F. Dziuba and R. E. Elmquist, "Improvements in resistance scaling at NIST using cryogenic current comparators," CPEM'92 Digest, pp. 284-285, June 1992.

[28] R. F. Dziuba and D. B. Sullivan, "Cryogenic direct current comparators and their applications," IEEE Trans. Mag., MAG-11, pp. 716-719, Mar. 1975.

[29] R. E. Elmquist and R. F. Dziuba, "Isolated ramping current sources for a cryogenic current comparator bridge," Rev. Sci. Instr., vol. 62, pp. 2457-2460, Oct. 1991.

[30] F. Delahaye, T. J. Witt, R. E. Elmquist, and R. F. Dziuba, "Comparison of quantum Hall effect resistance standards of the NIST and the BIPM," Metrologia, vol. 37, pp. 173-176, 2000.

[31] R. E. Elmquist and R. F. Dziuba, "Loading effects in resistance scaling," IEEE Trans. Instrum. Meas., vol. 46, pp. 322-324, 1997.

[32] A. Jeffery, R. E. Elmquist, and M. E. Cage, "Precision tests of a quantum Hall effect device DC equivalent circuit using double-series and triple-series connections," J. Res., Natl. Inst. Stand. Technol., vol. 100, pp. 677-685, 1995.

[33] B. N. Taylor and C. E. Kuyatt, "Guidelines for evaluating and expressing the uncertainty of NIST measurement results," Natl. Inst. Stand. Technol. Technical Note 1297, 1992.

[34] International Bureau of Weights and Measures (BIPM) Working Group on the Statement of Uncertainties, Metrologia, vol. 17, pp. 73-74, 1981.

[35] A. Jeffery, R. E. Elmquist, J. Q. Shields, L. H. Lee, M. E. Cage, S. H. Shields, and R. F. Dziuba, Determination of the von Klitzing constant and the fine-structure constant through a comparison of the quantized Hall resistance and the ohm derived from the NIST calculable capacitor, Metrologia, vol. 35, pp. 83-96, 1998.

[36] Introduction to Statistical Quality Control, 2nd Edition, Douglas C. Montgomery, John Wiley & Sons, New York, NY, pp. 232-233, (1991).

14. BIBLIOGRAPHY

1. Precision Measurement and Calibration: Electricity and Electronics, Natl. Bur. Stand. (U. S.) Handbook 77, vol. I, Feb. 1961.

2. Precision Measurement and Calibration: Electricity-Low Frequency, Ed. by F. L. Hermach and R. F. Dziuba, Natl. Bur. Stand. (U. S.) Spec. Publ. 300, vol. 3, Dec. 1968.

3. Precision Measurement and Calibration: Electricity, Ed. by A. O. McCoubrey, Natl. Bur. Stand. (U. S.) Spec. Publ. 705, Oct. 1985.

4. Electrical Measurements, F. K. Harris, John Wiley & Sons, New York, NY, 1952.

5. Precision DC Measurements and Standards, D. S. Luppold, Addison-Wesley, Reading, MA, 1969.

6. The Current Comparator, W. J. M. Moore and P. N. Miljanic, Peter Peregrinus Ltd., London, UK, 1988.

7. Quantized Hall Resistance Recommended Intrinsic-Derived Standards Practice, RISP-3, R. E. Elmquist, B. N. Wood, and K. Jaeger, Published by the National Conference of Standards Laboratories, Boulder, CO, Aug. 1997.

15. APPENDIX

15.1 Report for a 1 Ω "Thomas-type" Standard Resistor

15.2 Report for a 10 kΩ "Special Air-type" Standard Resistor

15.3 Report for a High-Value Standard Resistor

REPORT OF CALIBRATION

1 Ω Standard Resistor

Manufacturer
Model   , Serial Number

Submitted by:

This standard resistor was calibrated on March 1, 2004 under NIST Calibration 51130C. Its reported value of resistance is based on the results of measurements comparing this standard with NIST working standards calibrated in terms of the quantum Hall effect used as the U. S. representation of the ohm, as described in NIST Technical Note 1458.

Temperature (°C)	Pressure (kPa)	Resistance (Ω)	Uncertainty (ppm)
25.000	100.00	1.000 000 023	0.04

The reported uncertainty, consistent with practice recommended by the International Bureau of Weights and Measures (BIPM), is the expanded uncertainty, $U = k\, u_c$, where u_c is the relative combined standard uncertainty for all known sources of error and k is a coverage factor. Consistent with international practice, NIST uses a coverage factor of $k = 2$. The expanded uncertainty is thus given by

$$U = 2\, u_c = 2\sqrt{\sum u_i^2}$$

where u_i is the estimated relative standard uncertainty associated with the ith source of error and expressed as an equivalent standard deviation. One component is an estimated Type A standard uncertainty based on the pooled standard deviation from a large population of individual measurements. The calibration uncertainty is affected by the measurement parameters described on the attached fact sheet, and calculated using the method described in NIST Technical Note 1458. The reported uncertainty contains no allowances for the long-term drift of the standard under calibration or for the possible effects of transporting the standard between laboratories.

Measurements performed by: For the Director,

                                                                                  , Group Leader
Quantum Electrical Metrology Division   Quantum Electrical Metrology Division

Test Report No.:
Reference:
Date:
Telephone Contact:

STANDARD RESISTOR

Manufacturer
Model   , Serial Number

Submitted by:

This **FACT SHEET** applies to Thomas-type 1 Ω standard resistor calibrations

Measurement Parameters that are known to affect the calibration value for this type of resistance standard are temperature and pressure and their values at the time of calibration are presented on the calibration report.

Temperature:

The temperature of the oil bath in which the standard resistor is placed is a parameter that can be controlled so that its influence on the uncertainty of the resistance is negligible. In order for there to be a negligible effect on the uncertainty of this calibration, the temperature of the oil is controlled to be within ±0.003 °C of 25.000 °C.

Pressure:

Pressure is a measurement parameter that is not controlled during the calibration. This pressure value includes the ambient barometric pressure and the additional pressure effect on the resistor due to its depth of immersion in the oil. For this calibration the pressure is within ±0.1 kPa of the value listed on the report.

For further information concerning this calibration refer to NIST Technical Note 1458.

Test Report No.:
Reference:
Date:
Telephone Contact:

REPORT OF CALIBRATION

10 kΩ Standard Resistor

Manufacturer
Model , Serial Number

Submitted by:

This standard resistor was calibrated on March 1, 2004 under NIST Calibration 51131C. Its reported value of resistance is based on the results of measurements comparing this standard with NIST working standards calibrated in terms of the quantum Hall effect used as the U. S. representation of the ohm, as described in NIST Technical Note 1458.

Temperature (°C)	Pressure (kPa)	Resistance (Ω)	Uncertainty (ppm)
23.00	100.00	9 999.995 4	0.08

The reported uncertainty, consistent with practice recommended by the International Bureau of Weights and Measures (BIPM), is the expanded uncertainty, $U = k\, u_c$, where u_c is the relative combined standard uncertainty for all known sources of error and k is a coverage factor. Consistent with international practice, NIST uses a coverage factor of $k = 2$. The expanded uncertainty is thus given by

$$U = 2\, u_c = 2\sqrt{\sum u_i^2}$$

where u_i is the estimated relative standard uncertainty associated with the ith source of error and expressed as an equivalent standard deviation. One component is an estimated Type A standard uncertainty based on the pooled standard deviation from a large population of individual measurements. The calibration uncertainty is affected by the measurement parameters described on the attached fact sheet, and calculated using the method described in NIST Technical Note 1458. The reported uncertainty contains no allowances for the long-term drift of the standard under calibration or for the possible effects of transporting the standard between laboratories.

Measurements performed by: For the Director,

                                                             , Group Leader
Quantum Electrical Metrology Division   Quantum Electrical Metrology Division

Test Report No.:
Reference:
Date:
Telephone Contact:

STANDARD RESISTOR

Manufacturer
Model , Serial Number

Submitted by:

This **FACT SHEET** applies to special air-type 10 kΩ resistor calibrations

A measurement parameter that is known to affect the calibration value for this type of resistance standard is temperature, and the value of this parameter at the time of calibration is presented on the calibration report.

Temperature:

The temperature of the standard resistor in air is not controlled to the extent that its influence on the uncertainty of the resistance is negligible. For this calibration the temperature of the customer's standard as measured by a thermistor probe in its thermometer well is within ±0.05 °C of the temperature given on this report.

Pressure:

Pressure is a measurement parameter that is not controlled during the calibration. For this calibration the pressure is within ±0.1 kPa of the value listed on the report.

For further information concerning this calibration refer to NIST Technical Note 1458.

Test Report No.:
Reference:
Date:
Telephone Contact:

REPORT OF CALIBRATION

1 GΩ Standard Resistor

Manufacturer
Model   , Serial Number

Submitted by:

This standard resistor was calibrated on March 1, 2004 under NIST Calibration 51147C. Its reported value of resistance is based on the results of measurements comparing this standard with NIST working standards calibrated in terms of the quantum Hall effect used as the U. S. representation of the ohm, as described in NIST Technical Note 1458.

Temperature (°C)	Humidity (%)	Voltage (V)	Resistance (Ω)	Uncertainty (ppm)
23.00	35	100	$0.999\ 876 \times 10^9$	25

The reported uncertainty, consistent with practice recommended by the International Bureau of Weights and Measures (BIPM), is the expanded uncertainty, $U = k\ u_c$, where u_c is the relative combined standard uncertainty for all known sources of error and k is a coverage factor. Consistent with international practice, NIST uses a coverage factor of $k = 2$. The expanded uncertainty is thus given by

$$U = 2\ u_c = 2\sqrt{\sum u_i^2}$$

where u_i is the estimated relative standard uncertainty associated with the ith source of error and expressed as an equivalent standard deviation. One component is an estimated Type A standard uncertainty based on the pooled standard deviation from a large population of individual measurements. The calibration uncertainty is affected by the measurement parameters described on the attached fact sheet, and calculated using the method described in NIST Technical Note 1458. The reported uncertainty contains no allowances for the long-term drift of the standard under calibration or for the possible effects of transporting the standard between laboratories.

Measurements performed by: For the Director,

                                                     , Group Leader
Quantum Electrical Metrology Division   Quantum Electrical Metrology Division

Test Report No.:
Reference:
Date:
Telephone Contact:

STANDARD RESISTOR

Manufacturer
Model , Serial No.

Submitted by:

This **FACT SHEET** applies to standard resistors of value 10 MΩ and above.

Measurement Parameters that are known to affect the calibration value for this type of resistance standard are temperature, humidity, and the voltage used to calibrate the resistor. Values of these parameters at the time of the calibration are presented on the calibration report.

Temperature:

The temperature of a standard resistor in an air bath is a parameter that can be controlled so that its influence on the uncertainty of the resistance is negligible. In order for there to be a negligible effect on the uncertainty of this calibration, the temperature of the customer's standard is controlled to be within ±0.1 °C of 23.0 °C.

Humidity:

The humidity of the air in contact with the standard resistor is a parameter that can be controlled so that its influence on the uncertainty of the resistance is negligible. In order for there to be a negligible effect on the uncertainty of this calibration, the relative humidity is controlled to be within ±5% of 35%.

Voltage:

The measurement voltage applied to calibrate the standard resistor is a parameter that can be controlled so that its influence on the uncertainty of the resistance is negligible. In order for there to be a negligible effect on the uncertainty of this calibration, the voltage is controlled to be within 1% of the value given on the report.

For further information concerning this calibration refer to NIST Technical Note 1458.

Test Report No.:
Reference:
Date:
Telephone Contact:

www.ingramcontent.com/pod-product-compliance
Lightning Source LLC
Chambersburg PA
CBHW081847170526
45167CB00007B/2918